Binkley

Seeds of Life

By the Same Author:

John Taine

(Science Fiction)

The Purple Sapphire, 1924
Quayle's Invention, 1927
The Gold Tooth, 1927
Green Fire, 1928
The Greatest Adventure, 1929
The Iron Star, 1930
Before the Dawn, 1934
The Time Stream, 1946
The Forbidden Garden, 1947
The Cosmic Geoids, 1949
Seeds of Life, 1951

Eric Temple Bell

(Technical Books)

The Cyclotomic Quinary Quintic, 1912
An Arithmetical Theory of Certain Numerical Functions, 1915
Algebraic Arithmetic, 1927
The Development of Mathematics, 1940

(Quasi-Mathematical Books of General Interest)

Debunking Science, 1927
The Queen of the Sciences, 1931
Numerology, 1933
The Search for Truth, 1934
The Handmaiden of the Sciences, 1937
Men of Mathematics, 1937
Man and His Lifebelts, 1938
The Magic of Numbers, 1946
Mathematics, Queen and Servant of Science, 1951

JOHN TAINE

Seeds of Life

SCIENCE
FICTION

Decorations by Ric Binkley

FANTASY PRESS
READING, PENNSYLVANIA, 1951

FIRST EDITION

Printed in U. S. A.

CONTENTS

Seeds of Life

1

THE BLACK WIDOW

"DANGER. KEEP OUT." THIS CURT WARNING IN scarlet on the bright green steel door of the twenty million volt electric laboratory was intended for the curious public, not for the intrepid researchers, should one of the latter carelessly forget to lock the door after him.

The laboratory itself, a severe box of reinforced concrete, might have been mistaken by the casual visitor as a modern factory but for the fact that it had no windows. This was no mere whim of the erratic architect; certain experiments must be carried out by their own light or in the dim glow of carefully filtered illumination from artificial sources. The absence of windows gave the massive rectangular block a singularly forbidding aspect. An imaginative artist might have said the laboratory had a sinister appearance, and only a scientist would have contradicted him. To the daring workers who tamed the man-made lightnings in it, the twenty million volt laboratory was more austerely beautiful than the Parthenon in its prime.

Of the thousands who passed the laboratory daily on their way to or from work in the city of Seattle, perhaps a scant half dozen gave it so much as a passing glance. It was just another building, as barren of romance as a shoe factory. The charm of the Erickson Foundation for Electrical Research was not visible to a casual inspection. Nevertheless, its fascination was a vivid fact to the eighty men who slaved in its laboratories twelve or eighteen hours a day, regardless of all time clocks or other devices to coerce the unwilling to earn their wages. Their one trial was the fussy Director of the Foundation; work was a delight.

About three o'clock of a brilliant May afternoon, Andrew Crane and his technical assistant, the stocky Neils Bork, gingerly approached the forbidding door, carrying the last unit of Crane's latest invention. This was a massive cylinder of Jena glass, six feet long by three in diameter, open at one end and sealed at the other by an enormous metal cathode like a giant's helmet. It had cost the pair four months of unremitting labor and heart-breaking setbacks to perfect this evil-looking crown to Crane's masterpiece. Therefore, they proceeded cautiously, firmly planting both feet on one granite step leading to the green door before venturing to fumble for the next.

Their final tussle in the workshops of the Foundation had endured nineteen hours. The job of sealing the cathode to the glass had to be done at one spurt, or not at all. During all that grueling grind neither man had dared to turn aside from the blowpipe for a second. In the nervous tension of succeeding at last, they had not felt the lack of food, water, or sleep. They had failed too often already, each time with the prize but a few hours ahead of them, to lose it again for a cup of water.

Crane planted his right foot firmly on the last, broad step. His left followed. He was up, his arms trembling from exhaustion. Bork cautiously felt for the top step. Then abused nature took her sardonic revenge for nineteen hours' flouting of her rights. The groping foot failed to clear the granite angle by a quarter of an inch. In a fraction of a second, four months' agonizing labor was as if it had never been.

Crane was a tall, lean Texan, of about twenty-seven, desiccated, with a long, cadaverous face and a constant dry grin about his mouth. He shunned unnecessary speech, except when a tube or valve suddenly burnt itself out owing to some oversight of his own. When Bork blundered, Crane as a rule held his tongue. But he grinned. Bork wished at such awkward moments that the lank Texan would at least swear. He never did; he merely smiled. Neils Bork was a true Nordic type, blue-eyed, yellow-haired, stockily built. From his physical appearance he should have been a steady, self-reliant technician. Unfortunately he was not as reliable as he might have been, had he given himself half a chance.

Viewing the shattered glass and the elaborate cathode, which had skipped merrily down the granite steps, and was now lying like a capsized turtle on its cracked back thirty feet away in the middle of the cement sidewalk, Crane grinned. Bork tried not to look at his companion's face. He failed miserably.

"I couldn't help it," he blurted out.

It was a foolish thing to have said. Of course he couldn't 'help it'—now. Only an imbecile would have deliberately smashed an intricate piece of apparatus that had taken months of sweating toil to perfect. Bork's indiscretion loosened Crane's reluctant tongue.

"You could help it," he snapped, "if you'd let the booze alone. Look at me! I'm as steady as a rock. You're shaking all over. Cut it out after this, or I'll cut you out."

"I haven't touched a drop for—" the wretched Bork began in self-defense, but Crane cut him short.

"Twenty hours! I smelt your breath when you came to work yesterday morning. Of course you can't control your legs when you're half-stewed all the time."

"It was working all night that made me trip. If you had let me take a layoff after we finished, as I asked, instead of carrying it over at once, I wouldn't—"

"All right. Keep your shirt on. Sorry I rode you. Well," Crane continued with a sour grin, "we shall have to do it again.

That's all. I'm going to take a look around before I go to bed. Let's see if the baby can still kick."

Bork stood wretchedly silent while Crane unlocked the steel door.

"Coming?" Crane called, as he switched on the lights.

Bork followed, locked the door, and stood sullenly beside his chief on the narrow steel gallery overlooking the vast pit of the huge transformers. Forty of these towering giants, gray and evil as the smokestacks of an old time battleship, loomed up menacingly in the glaring light. Each stood firmly planted on its towering tripod—three twenty-foot rigid legs made up of huge mushroom insulators, like a living but immobile enemy from another planet. The whole battery of the forty devils presented a strangely half human aspect, and their massed company conveyed a sinister threat, as of seething whirlwinds of energy stored up against the men who had rashly created these hostile fiends. The two men, staring down on their half-tamed genii, felt something of this menace, although both were practical and one was daring almost to a fault. But in their present exhaustion, nature succeeded in making herself felt, if not heard, on a deeper, more intuitive level of their consciousness.

"Let's try out the two million volt baby," Crane proposed as a peace offering to the still surly Bork. "We haven't busted that, yet," he continued rather tactlessly, and Bork shot him a spiteful glance.

The 'two million volt' to which Crane referred was his first attempt to build a more powerful X-ray tube than any then in existence. By studying this two million volt baby minutely, Crane hoped to succeed with the full grown twenty million volt tube which he and Bork were constructing. Then, if theory for once should prove a trustworthy guide to the riddle of matter, they hoped to smash up the atoms of at least half a dozen of the elements. What might happen thereafter Crane refused to predict. He had seen too many ingenious theories exploded suddenly and finally by some unforeseen 'accident,' to have much faith in prophecies not founded on experimental evidence.

"Step it up two hundred and fifty thousand at a time," he ordered Bork, "and be careful. We don't want to blow out the tube."

Again Bork shot him a resentful glance as if his bruised conscience accused him of being a hopeless bungler. Nothing was farther from Crane's mind. He was merely repeating the routine instructions of the laboratory. To prevent possibly fatal mishaps, the experimenters invariably followed a rigid set of rules in their work, testing every switch and piece of apparatus in a definite order before touching anything, although they 'knew' that everything was safe.

Bork threw in the first switch and turned off the lights, plunging the laboratory into total darkness. There was a metallic clang, and the black air began to vibrate ominously with a rapid, surging hiss. A sombre eye of cherry red stole out on the darkness as two hundred and fifty thousand volts flashed to the cathode of the X-ray tube; then, almost instantly, the red flashed up to a dazzling white spot.

"All right," Crane ordered, "throw in the next."

Under half a million volts the twelve foot tube flickered and burned with a fitful green fluorescence, revealing the eight metal 'doughnuts,' like huge balloon tires, encircling the glass. These constituted the practical detail which balanced the terrific forces within the tube and prevented the glass from collapsing.

The outcome of any particular 'run' was always somewhat of a sporting venture. Until the shot was safely over, it was stupid to bet that the tube would not collapse or burn out. Crane waited a full two minutes before ordering the step up to seven hundred and fifty thousand volts. Under the increased pressure the surging dry hiss leapt up, shriller and angrier, and deep violet coronas of electricity bristled out, crackling evilly, in unexpected spots of the darkness.

Bork began to grow restless.

"Hadn't we better step it up to a million now, and quit?"

Crane laughed his dry laugh. "Getting nervous about what Dr. Brown told us?"

Bork grunted, and Crane, in his cocksure ignorance, elucidated. "All doctors are old women. What do the physiologists actually know about the effect of X-rays as hard as ours on human tissues? I've spent at least thirty hours the past eight months working around that tube going at capacity—two million volts—and there isn't a blister or a burn anywhere on my body. I'll bet these rays are so hard they go straight through flesh, bone and marrow like sunlight through a soap bubble. What are you afraid of? If our bodies are so transparent to these hard rays that they stop none of the vibrations, I fail to see how the biggest cells in us are in any danger whatever. You've got to stop hard radiations, or at least damp them down, before they can do human bone, nerves or muscles any harm. All the early workers used soft rays. That's why they lost their eyesight, fingers, hands, arms, legs, and finally their lives."

"It takes months, or even years, for the bad burns to show up," Bork objected.

"Well," Crane retorted, "if there is anything in what Dr. Brown said, I should be a pretty ugly leper right now. Use your eyes. My skin's as smooth as a baby's."

"He said you will be sterilized for life," Bork muttered. "The same for me. I'm not going to live the next twenty years like a rotten half-man."

"Be a confirmed bachelor like me," Crane laughed, "and you'll never miss the difference. What's a family anyway but a lot of grief? Throw in the next switch and forget the girl."

Under the million volts, the glowing tube buzzed like a swarm of enraged hornets, and for the first time in all his months of work in the laboratory, Crane felt a peculiar dry itching over his whole body. As Bork stepped the voltage up to the full two million, the itching increased to the limit of endurance.

"Imagination," he muttered, refusing to heed nature's plain hint. "Hand me the fluoroscope, will you?"

Bork groped over the bench beneath the switches and failed, in the dark, to find what he sought.

"I'll have to turn on the lights."

"Very well. Make it snappy. I need my lunch and a nap. So do you."

Rather than admit that Bork's fears might not be wholly old-womanish, Crane would stick out his discomfort and delude his assistant into a false feeling of security by feigning an interest in the hardness of the rays.

Bork turned on the floodlights. Just as he was about to pick up the fluoroscope, he started back with an involuntary exclamation of disgust. His arm shot to his side as if jolted by a sharp shock.

"Short circuit?" Crane snapped. "Here, I'll pull the switches."

In two seconds the coronas were extinct, a succession of metallic clanks shot rapidly to silence, and the cathode of the two million volt tube dimmed to a luminous blood red. The tingling itch, however, on every inch of Crane's skin persisted. Bork for the moment was apparently beyond speech. In the glaring light his face had a greenish hue, as if he were about to be violently seasick.

"Short circuit?" Crane repeated.

"No," Bork gasped. "Black widow."

Crane failed to conceal his contempt. "Afraid of a spider? Why didn't you smash it?"

Bork swallowed hard before replying.

"It dropped off the bench and fell behind those boards."

"Rot! You're seeing things. It'll be snakes next. There have been no black widows found nearer than Magnolia Bluffs or Bainbridge Island—ten miles from here."

Crane's indifferent sarcasm stung Bork to cold fury. His nerves were undoubtedly on edge after nineteen hours' exasperating work and months of more or less steady, 'moderate' soaking. He succeeded in keeping his voice level.

"Snakes? Then lift that board."

Without a word, Crane bent down and contemptuously tossed the top board aside. "There's nothing here," he remarked dryly, turning the next board. In his zeal to discomfit Bork he deliberately thrust his hand into the narrow space between the pile of boards and the wall, sweeping it methodically back and forth to dislodge the supposedly imaginary enemy. The sweat started out on Bork's forehead. Death by the bite of an aggressively venomous spider is likely to be unpleasant even to witness.

"Look out!" Bork yelled, as a jet black ball, the size of a tiny mouse, rolled from behind the pile, instantly took energetic legs to itself, and scurried with incredible speed straight up the concrete wall directly before Crane's face. Crane's action was instinctive. He straightened instantly to his full height, gave a convulsive leap and, with his clenched fist, smashed the loathsome thing just as it was about to scud beyond his reach. It fell, a mashed blob of evil black body and twitching legs, plop into the eyepiece of the fluoroscope.

"You win this time," Crane grinned, turning the black mess over on its back. "She's a black widow. Here's her trademark—the red hour glass on her underside. We had better post a warning to the fellows to go easy in the dark. This is the ideal breeding place for the brutes—dry and warm, with plenty of old packing cases lying about. I'll have to ask Mr. Kent to get this cluttered rats' nest cleaned up for once. Well, shall we finish our shot?"

"What for?" Bork demanded.

"Just to prove that we haven't lost our nerve. Here, I'll remove the evidence from the fluoroscope before you douse the lights. Better save the remains for the Director," he continued, carefully depositing the smashed spider in an empty cigar box, "or he'll say we've both been hitting the bottle. Ready? Shoot; I've got the fluoroscope."

As the lights went off, Crane caught the dull flash of anger on Bork's face. "I had better stop prodding him," he thought, "or he may stick a knife into me. He's a grouch; no sense of humor."

Crane was partly right. Bork, a poorly educated mechanic with a natural gift for delicate work, cherished a sour grudge against the world in general and against the eighty trained scientists of the Erickson Foundation in particular. They, he imagined, had profited by the undue advantages of their social position, and had somehow—in what particular way he could not define—swindled him out of the education he merited. He had been denied the fair opportunity, which a democracy is alleged to offer all comers, of making something of himself. Such was his aggrieved creed.

As a matter of fact a good third of all the scientists on the staff had earned their half starved way through high school, college and university with no greater resources at their command than Bork possessed when he was at the student age. That they preferred drudgery for a spell to boozy good fellowship for the term of their apprenticeship accounted for the present difference between their status and his. One of these men, a great specialist in X-ray crystal analysis, had paid his way while a student by stoking coal eight hours a night in the municipal gas plant. Bork, in all his flaming youth, had done nothing more strenuous than act as half time assistant, four hours a day, to a pattern maker.

Bork had brains; there was no denying so obvious a fact. But he was short on backbone. Being Crane's technical assistant, he naturally, if only half consciously, stored up all his spite against life for Crane's special amusement. Crane was the one man in the Foundation who could have tolerated the grouchy

Bork for more than a week. The rest would have discharged him without compunction. Crane's wry sense of humor gave him a more human angle on the dour churl. Although he would have cut his tongue out, rather than acknowledge the fact, even to himself, Crane hoped to save Bork from his sourer fraction and make a man of him. This missionary drive lay behind his frequent digs at Bork's tippling. Crane sensed the man's innate ability. That all this high grade brain power should fritter itself away on peevish discontent and sodden conviviality seemed to him an outrage against nature.

The exasperations of this particular day, culminating in the wreck of the new cathode and the incident of the black widow, crystallized Bork's sullen irritation toward Crane into a definite, hard hatred. The uninitiated often marvel at the trivial grounds cited by the injured party in a divorce suit, overlooking the ten or fifteen years of constant fault-finding and mutual dislike concealed beneath the last, insignificant straw.

So it proved in Bork's case. Crane's superior contempt for his assistant's perfectly natural abhorrence of a venomous spider revealed the full measure of the stronger man's subconscious scorn for a weakling. Bork was no fool. He realized that although Crane had always looked down on him as a somewhat spineless parody of a full grown man, he himself had looked up to Crane, not with respect or affection, but with smouldering hatred and the unacknowledged desire to humble the better man to his own pygmy stature. And in that sudden flash of revelation, struck out on the darkness of his thwarted nature by a tactless jest, Bork saw himself as the appointed destroyer of his would-be friend and natural enemy. His bitter sense of inferiority was swallowed up in a yet more bitter certainty that his was the power to injure Crane in a way that would hurt.

As he switched off the floodlights, and silently threw in the full two million volts in eight perfectly timed steps of two hundred and fifty thousand each, he resolved to get blind drunk the moment he was free of Crane's supervision. He would not dull the edge of his projected spree by foolishly indulging in lunch or supper. No; he would hurl himself and all his forces raging and ravenously empty on the crudest, rawest Scotch whiskey he could buy. What should happen thereafter would be up to Crane alone. In any event Bork would win, in his perverse way, even if it cost him a term in the penitentiary.

"How's that for penetration?" Crane demanded enthusiastically, holding his hand before the fluoroscope in the path of the rays. They were standing about a hundred feet away from the tube. Not a shadow of flesh or bone showed on the fluoroscope. To those hard rays, the human body was as transparent as rock crystal to sunlight. Bork gave a grudging consent that it was pretty good. To test the penetration further, Crane next tried to cast a shadow of the heavy iron rail, against which he was leaning, on the fluorescent screen. Again the penetrating radiation passed clear through the obstacle as if it were air.

"And you're afraid," Crane exulted, "that rays which will pass like these through iron can affect the insignificant cells of your body. They wouldn't bother to stop for such stuff." Nevertheless it cost Crane all of his self control to keep from tearing at his own tingling, itching skin.

"Well, let's call it a day, and go home," he said.

On emerging from the laboratory they found a knot of curious idlers gathered about the cracked cathode, vainly trying to puzzle out whose the huge 'helmet' might be.

"We had better rescue that," Crane remarked, "before some loafer finds out that it's valuable. We can't afford to lose several hundred dollars' worth of platinum on top of our other hard luck."

Crane's thoughtless allusion to their mishap was the last straw. With a smothered oath, Bork turned his back on the small crowd and strode off toward the street.

"See you tomorrow at eight," Crane called after him.

Bork made no reply. Grinning broadly, Crane picked up the cathode and started with it back to the workshops. The idlers, having thoughtfully selected choice souvenirs of broken glass, dispersed. Had Crane been as keen a student of human nature as he was of the physics of radiation, he would have followed Bork and let the crowd keep the costly cathode as a memento of a memorable blunder.

2

THE BOILING BOX

INSTEAD OF HASTILY SWALLOWING A MEAL and hurrying home to bed as he had intended earlier in the afternoon, Crane sped as fast as his long legs would take him to see his physician.

Dr. Brown, the specialist in radiology, who had already warned Crane of the possible consequences of exposing himself recklessly to the hard X-rays, lived within a quarter of a mile of the Erickson Foundation. Being a family physician to about half the staff of the Foundation, he understood their needs better than might the average doctor. More than once he had been called out of bed in the small hours of the morning to resuscitate some careless worker who had neglected the precautions of common sense and been jolted into insensibility, or to pick splinters of glass from hands and faces damaged in the pursuit of science. Brown himself specialized in medical radiology, and was expert on everything that an up-to-date physician should know about the action of cosmic rays, X-rays, and ultra violet light on the human body. As a hobby he kept abreast of biology in its less practical phases, particularly in a study of the protozoa.

Crane found the doctor in. Without preliminaries of any

kind, he plunged into the middle of things.

"My whole skin burns and itches like the very devil."

"You've been working with your two million volt tube again? Without any protection, as usual?"

Crane nodded, extending his bare forearm for Brown to examine. The doctor studied the skin minutely through a powerful pocket lens and shook his head.

"If there's anything wrong, a microscopical examination of the skin may show it up. Everything looks perfectly normal through this. Sure it's not just your imagination running away with what I said the other day?"

For answer Crane, unable longer to control himself, began tearing with his nails at every accessible inch of skin on his body. Brown rose and filled his hypodermic.

"This will stop it for a time. Go home and take a starch bath. Then rub down with calamine ointment. If the itch comes back, stick it out as long as you can before calling me. I'll probably be at home all evening. If not, the housekeeper will give you the name of another man. He will know what to do."

As the powerful shot took effect, the intolerable itching became bearable, and Crane began to doubt that his discomfort was more than an attack of nerves. Nevertheless he carried out the doctor's instructions to the letter.

"Safety first," he grinned, stepping from the milky starch bath and reaching for the towel. In his eagerness to live up to the doctor's orders, Crane hastened to dry himself thoroughly and rub down his whole body with calamine before even pulling the plug of the bath tub. Having finished his rub, he turned round to let the water out, and stopped short with an exclamation of amazement. The water, milky white less than five minutes before, was now a vivid pink. Even as he watched, the color deepened from red to crimson. In ten seconds the strange fluid had taken on the characteristic hue of freshly shed blood. Crane flung on his bathrobe and ran to the telephone.

In his haste to call Doctor Brown, Crane forgot to shut the bathroom door after him. His landlady chanced to pass along the corridor on her way down to the kitchen, just as Crane, in the telephone alcove, took down the receiver. Like the good housekeeper she was, the landlady made a move to close the bathroom door on her way past. The bathtub full, apparently, of human blood, paralyzed her for two seconds before she screamed. As she fled shrieking from the house, Crane succeeded in getting his connection. Doctor Brown, listening at the other end of the wire, heard ear-splitting shrieks and a man's voice which he failed to recognize as his patient's requesting him to come at once to Crane's apartment. He banged the receiver back on the hook and grabbed his emergency kit. Crane, he imagined, driven insane by his torments, had attempted to commit suicide.

On reaching Crane's apartment house, the doctor ran slap into enough excitement to justify a dozen murders and suicides. The landlady, in hysterics, was being supported on the lawn by two sympathetic neighbors. Crane, gorgeous in a flaming orange bathrobe which flapped about his long legs, was doing his best to convince three motorcycle policemen and a clamoring mob of morbid sensation hunters that he had committed no murder, but had merely indulged in a late afternoon bath. The police had their hands full keeping the mob back.

With the skill acquired from many adventures with crowds and accidents, Doctor Brown insinuated himself into the mob and quickly worked his way to the police.

"I'm the doctor they telephoned for. What's up?"

"Nothing," the officer replied disgustedly, "if that fellow in the kimono knows what he's talking about. Go in and 'phone headquarters to send half a dozen men to help us."

Brown joined Crane on the porch, snatched him into the house, and bolted the door. Then he telephoned to the police.

"What happened?" he demanded of Crane, on receiving the Chief's assurance that the riot squad was on its way.

"I followed your instructions," Crane grinned. "Come and have a look at the bathtub."

With dramatic effect, Crane ushered the doctor into the bathroom and gestured toward the tub. Then his jaw dropped. The water was as starchy white as when he had stepped from the tub. Not a trace of all that violent blood remained.

"Well?" the doctor demanded meaningly.

"The landlady saw it too," Crane began. "I'm not crazy."

"Saw what?"

Rather shamefacedly, Crane gave a short, but complete account of the entire incident as it had seemed to happen. To his surprise, Brown did not laugh.

"You think there may be something in it?" Crane ventured.

Brown was non-committal. He suspected Crane of a nervous breakdown, but refrained from saying so. The landlady doubtless had been scared half out of her senses by some stupid practical joke on Crane's part.

"Let me take your temperature."

Crane submitted. His temperature was normal. So, as far as Brown could judge, was everything else about him. The theory of a nervous breakdown was abandoned.

"Find me a clean, empty bottle or a jam jar. I'll take a sample of the water and find out if there is anything wrong with it."

While Crane rummaged in the kitchen, the doctor carefully salvaged the teaspoonful of starch remaining in the empty cardboard container. He was just conveying this to his bag, when the front door bell began ringing insistently. At the same instant Crane reappeared with a clean ketchup bottle.

"Don't answer the bell till I fill this. Otherwise the police may smell a rat and bring the reporters down on our necks."

Hastily stowing the bottle of starchy water into his handbag, Brown followed Crane to the front door. The instant the bolt was drawn, a hard-faced captain of detectives thrust himself into the hallway.

"Where's the bathroom? You show me," he suggested grimly, seizing Crane's arm.

"Sure," Crane grinned. "The city waterworks, if you like."

Deigning no reply, the captain hustled the suspect upstairs. Once in the bathroom he gave a disgusted grunt at the tub, picked up the rag rug, scrutinized it thoroughly, and finally inspected the articles in the toilet cabinet.

"Does your landlady drink?" he demanded sourly.

"Never touched a drop in her life," Crane gallantly assured him.

"Then she's bughouse. If she throws another party like this one, she goes to the asylum. Tell her that from me."

Turning on his heel, he quit the profitless investigation and clumped downstairs. In his cocksureness that he saw through everything, he overlooked the one clue of any value. It did not enter his head to quiz the doctor then waiting in the hallway till the quieted mob should disperse. What was a doctor doing in the house if everything was as it seemed to be? Who had called him? Why? For failing to think of these pertinent questions the skeptical captain deserved to lose at least one stripe. Brown, for his part, tried to make himself and his telltale black bag as inconspicuous as possible. He might have saved himself the trouble; his estimate of the captain's intelligence was far too high.

The moment the front door closed on the redoubtable captain, Brown darted for the stairs. He met Crane half way.

"If your skin starts itching again, come to my house at once. I'll give you a bath. Tell the landlady the heat affected her. I'll speak to her on the way out."

"You think——" Crange began.

"Nothing. But it will be worth while to analyze this water, or whatever it is. Well, I'm going before the reporters arrive. They will have got wind of this at the police station. If anyone asks you anything, leave me out of it. I can't afford this kind of advertising."

By the time Crane was dressed, the disappointed mob had dispersed, and the distraught landlady was doing her feeble best to fend off the persistent attacks of three able young reporters. Crane routed them.

"Beat it," he ordered curtly, entering the living room. "Can't you see that this lady is suffering from the heat? That's all there is to it. If you can make a story out of that you beat Hearst. Only," he concluded with a grin as he bowed them out of the house, "the city editor will scrap what you write. It would be a slam at our beautiful climate. There's a dog fight down the street. Try your luck on that. Scat!"

Having disposed of the press, Crane returned to the sitting room to comfort the landlady.

"Is there anything I can get you?" he asked sympathetically.

"If you don't mind, you might bring me a little gin and water—not too much water. The bottle is on the top shelf of the kitchen cabinet."

The last information was superfluous so far as Crane was concerned. He had discovered the half empty bottle while rummaging for what the doctor wanted. He was careful not to ruin the landlady's pick-me-up by too much water. In fact he gave her half a tumberful straight, which was just what she needed. As she sipped the fiery stimulant, the poor woman felt as guilty as sin. She resolved to make the present bottle her last. Bathtubs full of sudden blood are too high a price to pay for a quart a day. To her credit, she lived up to her resolution. This was a pity, as gin afforded the poor woman her one escape from her humdrum existence, and she had a constitution that sulphuric acid could not have corroded. When Crane learned of her self-denial he felt quite conscience-stricken. But he dared not tell her that her vision was a sober fact. Her first act would have been to eject him from his rooms, as an incurable of some particularly dangerous kind. Then, to square herself in the eyes of her neighbors, she would have confided the whole gruesome truth to the avid press.

Having seen to the landlady's comfort, Crane attended to his own. He slipped out to a restaurant, had a square meal, and hurried back to bed. By eight o'clock he was between the sheets, determined to sleep in spite of the faint prickling all over his body. It was beginning again gently; wondering whether he could cheat the enemy by falling asleep, Crane dozed off. He had won, for the time being.

The moment he had finished his dinner, Doctor Brown settled down to analyze the starchy water. Not being a skilled chemist he had to try the only method in which he was expert—microscopic examination. Should this reveal nothing unusual, he would submit a sample of the water to a competent chemist for detailed analysis.

Brown approached his problem with an open mind. Having profited by an excellent medical training, he was not addicted to forming opinions in advance of the evidence. There might be precisely nothing strange about that starchy water, or it might give a patient observer his first glimpse of a new universe. The doctor was ready for either contingency, or for neither.

With matter of fact deliberation he prepared the slide and carefully adjusted the microscope. Long years of habit enabled him to come within a shade of the true focus without looking through the eyepieces. His instrument was a high-powered binocular, which threw up the minute objects on the slide into solid relief. Having adjusted the focus roughly as far as was safe, Brown completed the delicate operation while looking through the eyepieces.

The singularly beautiful starch grains swam into stereoscopic view and blurred out, as the slowly moving lens sought to bring up the light from still minuter specks within reach of human vision. On the extreme threshold of visibility, a new universe slowly dawned.

The silent watcher of that undiscovered heavens scarcely breathed. Hour after hour he sat entranced, far from this world, as the first acts of a titanic drama, never before imagined by the human mind, unrolled majestically in a drop of water so tiny that no unaided eye could see it.

While one man followed the dim beginnings of a new order in wonder and awe that fateful night, another, blind with hatred and ignorant of what he was doing, sought to destroy the instrument of fate which had thrust the unknown universe up to the light and life. True to his brainless vow, Bork got soddenly drunk as fast as bad whiskey would let him. A dispassionate critic from a wiser planet, if confronted with Brown, Crane and Bork, that evening, might well have doubted that the three were animals of one and same species—*homo sapiens,* so-called. Brown and Crane he might easily have classified as sports from the same family tree; Bork undoubtedly would have puzzled him.

The vile whiskey, more like crude varnish and alcohol than a civilized drink, had an unprecedented effect on Bork's exhausted body. His customary experience after a full quart of the stuff was a feeling of general well being and a greatly inflated self-esteem. This spree was strangely different from all of its innumerable predecessors. Instead of experiencing the comforting glow which he anticipated, the wretched man felt chilled to the bone. The world turned gray before his eyes. He saw his own life, pale and ineffectual as a defeated spirit, wandering aimlessly hither and thither through a cheerless fog of unending years. Why must he endure it longer? The drink had made him deadly sober. A remorseless tongue, loosened by the alcohol, insinuated that he was an unnecessary accident, a dismal thing that should never have been trusted with life, and a mistake to be corrected and forgotten as quickly as possible. Crane was right, the unhappy man admitted although he had never put his estimate into cold, precise perspective, as Bork's own silent accuser was now doing. The very unconsciousness of Crane's contempt was what stung.

Bork sighed, slowly extracted the cork from the second quart, poured out a tumberful of the raw stuff, and sat meditatively sipping it. As the alcohol seeped into his tissues, the insane clarity of his mind increased.

By midnight he had disposed of three full quarts of alleged whiskey. This equalled his previous record. By one in the morning he had bettered his record by a quart and was ripe for his insanely logical action. In his normal state he could not have reasoned so coldly, so clearly, so consistently. Crazed by the drink, he became as rigidly logical as a hopeless maniac. Glancing at his watch, he rose steadily to his feet, and marched from his frowsy lodging to execute his purpose and to silence that persistent voice which said he was born a fool.

No one meeting him on the street could have guessed that he was not cold sober. That is exactly what he was. For the first time in his life Bork had discovered himself, completely and without the slightest reservations of false self-respect. He proceeded through the night with a firm, resolute step, like a plucky man going to the electric chair and determined to die game.

It was a few minutes past two in the morning when Bork silently let himself into the twenty million volt laboratory and locked the door behind him. His immediate business was too serious to admit a smile. Yet he almost smiled as he reflected that Crane would have the satisfaction of thinking—not saying—"that fool Bork has blundered again."

Crane's exasperating silence had given Bork the first inkling of the bitter truth. Now he, the despised assistant, the man who might easily have been chief had nature not loaded the dice against him, would prove that he was the better man in one thing at least—unanswerable destruction. Crane, the arrogant, strong man, whose vice was pride in his easy strength, should learn the meaning of frustration.

The man who can hurt his friend is, after all, the stronger of the two. Bork switched on the floodlights and prepared to prove his superior strength.

He set about his awful business deliberately, determined for once not to blunder. Having descended the steel stairs to the vast pit of the transformers, he made the necessary connections to link up the forty huge gray devils into a single unit. The forty were now ready to smite as one, with the full bolt of their twenty million volts, whatever accident or design might offer them to destroy. Like a callous priest preparing a peace offering to Moloch, Bork quickly and accurately made ready the pride of Crane's heart as a sacrifice to the forty united devils. The two million volt X-ray tube, Crane's "baby," was linked into the chain of destruction to appease the forty and prove Bork a better man than his tormentor.

His preparation was not yet complete. Not only must Crane be humbled, but the pitiless logic of his own subconscious mind must be refuted forever. He connected one end of a long, stout copper wire to that which was to feed twenty million volts into the X-ray tube, made a large loop of the free end, passed the loop over his left arm, and dragged the trailing wire after him up the steel steps to the gallery of the switches.

To secure the effect he wanted, all switches must be closed simultaneously, releasing twenty million volts in one flash. This presented no difficulty. The switches he must use were ranged in one horizontal line eight feet long. Temporarily winding the loop round an iron upright, Bork was ready for his problem.

First he all but closed the whole row of switches, bringing the eight foot line of ebonite handles into the same sloping plane. Then he glanced about for a narrow eight-foot board or strip. The pile of scrap lumber, behind which Crane had thrust his hand to scare out the black widow, contained just what Bork sought. At the bottom of the pile were several narrow lengths of white pine, the remains of large packing cases, from six to twelve feet long and four to six inches wide.

Taking no chance of encountering another venomous spider, Bork disengaged the desired piece of lumber with his foot before venturing to pick it up. Then, ashamed of his lack of courage in the face of what he was about to do, he propped the narrow board against the iron railing and turned to the bench beneath the switches. To prove himself not an utter coward, he put forth a steady hand and raised the lid of the cigar box into which Crane had dropped the crushed remains of the black widow.

She still lay there, black and venomous looking as death itself. Every fiber of her apparently was dead. As he stared down in cold fascination at the hideous, crushed thing, Bork detected not the slightest tremor in any of her eight long, smooth, black legs. All were curved stiffly inward, rigid in death, above the red hour glass on her mangled body. She was dead. To put his sorry manhood to a crucial test before joining the thing he instinctively loathed in death, Bork put out a finger and lifted each leg in turn. Six remained attached to the body, two dropped off.

"Dead," he muttered, closing the lid of the cigar box to shut out the sight of that repulsive corpse which he no longer feared worse than death itself.

His final preparations were brief. In a few seconds he had the copper loop fastened securely about his neck, and the long narrow board evenly balanced in both hands. He laid the board lightly along the eight foot row of ebonite switches. Then, with a convulsive movement of both arms, he shoved the board back and down, instantly throwing in the full battery of switches and releasing the irresistible fury of twenty million volts to shatter everything in their fiery path, himself included, to a chaos of incandescent atoms.

The instantaneous surge of energy missed one of its marks.

A deafening report followed the blinding green flash where the copper junction of the wire which was to have electrocuted Bork vaporized instantly. Before the current could leap along its entire length the first twenty feet of the wire exploded to atoms in a cloud of green fire. Bork's efficiency in making the connection had saved his life; a looser contact would have let but a fraction of the current through the wire to destroy him before it consumed the conductor. He had blundered again. His grandiose project had nullified itself in a short circuit which he should have foreseen.

Dazed and uncertain whether he was still living, he stared uncomprehendingly over the pit of the transformers and upon the X-ray tube. The floodlights, on an independent circuit, still filled the laboratory with an intolerable glare. Not a trace of corona flickered on any of the apparatus. The giant transformers loomed up, cold, gray and dead. The echo of the exploded wire seemed still to haunt the oppressive silence.

Gradually the stunned man became aware of the X-ray tube. Built to withstand the impact of two million volts, it should have been annihilated under the surging shock of twenty million. Had it taken the full bolt, or had the half foot of wire from the cathode to Bork's too efficient connection volatilized before the current could leap the short gap? That it received at least a fraction of the intended maximum was evident, for the lower half of the tube quivered and scintillated in coruscating pulses of sheer white light. The upper half of the vacuum in the tube, from the cathode down, was as black as ebony. Impenetrable darkness and sheer light were severed from one another absolutely; no shadow from the black dimmed the upper brilliance of the seething light, and no pulse of the white fire greyed the massive black above the invisible barrier.

Whatever might be taking place in that tube, it was automatic and independent of any extraneous electrical influences. The wires connecting the tube to the feeding apparatus had burned out, and the entire laboratory, except the floodlights, was electrically dead. As Bork watched, the black crept slowly downward. The diminishing light, devoured from above by the descending void, increased in intensity, as if struggling fiercely to resist and vanquish the death which crept down upon it. To the dazed man it appeared almost as if a plunging piston of black steel were compressing the resistant light down to nothing.

Within three minutes but half an inch of dazzling white fire remained. Laboring against the last desperate struggle of the light to survive, the black crept down more slowly. The last half inch dwindled to a mere plane of light as fiercely brilliant as the furnace core of a star. Then, in a second, the last light vanished, and the tube, now wholly black, exploded with a report that rocked the laboratory like an earthquake and hurled Bork to the steel floor of the gallery.

When he came to his senses, he found himself staring up through a phosphorescent glow to the dimly visible concrete ceiling. The explosion which had stunned him had shattered the globes of the floodlights. He got to his feet and reeled toward the door, only to trip over the copper wire dangling from his neck. With a curse he freed himself and fumbled for his key. Some moving object impinged gently against the back of his hand, seemed to break silently, and dispersed, leaving behind it only a faint sensation of cold. Another struck him in the face, and again he sensed the outward flow of heat from his skin. He became aware that his hands and face were slowly freezing.

To escape from that silent place of horror was his one instinct. The key in his pocket eluded his clumsy, half frozen fingers. Still dazed, he did not seek to discover the source of those moving things that touched his bare face as gently as kisses in a dream and slowly drained his body of its natural heat. At last he managed to grasp the key and insert it in the lock. His chilled fingers refused to function. He began beating wildly with his numbed fists against the steel door, conscious that he was slowly dying. Almost together, two of the moving objects softly struck the steel above his head, lingered for a moment and vanished into total darkness. He saw what they were.

The black air of the laboratory was alive with thousands of spinning vortices of faint light drifting in all directions, rebounding unharmed from one another when two or more collided, and dying only when they struck some material obstruction—walls, ceilings or apparatus. It was the mazy wanderings of this silent host which revealed the darkness against flickers and flashes of dim, tumultuous light. Their numbers diminished rapidly, for they seemed to seek their own extinction, quickening their motion as they drew near to solid substances and jostling one another in their eagerness to cease to be. In ten minutes the darkness would have conquered, but Bork did not wait to see its victory.

A slight rustling on the bench behind him made him spin round in anticipatory fear. Almost before the horror happened he sensed its advent. The lid of the cigar box, in which the crushed black widow lay, flipped up as if some frantic living thing were trying to escape. The lid subsided for an instant, then again flapped sharply up an eighth of an inch. Bork reached the door in one leap. This time his fingers functioned automatically. Glancing back as he flung open the steel door, he saw in the dim phosphorescence of the expiring vortices, a sight that reached the very roots of his fear.

The lid of the cigar box was thrown completely back on the bench by a rapidly swelling black mass that foamed up explosively from the box like living soot. As he slammed the door with a reverberating clang he caught a last glimpse of the boiling black mass budding upon itself in furious vitality and overflowing bench, platform and stairs in one hideous deluge of unnatural life.

He turned the key in the lock and reeled off into the icy grey pearl of the stirring dawn, sane at last with an awful sanity such as he had never known.

3

REBORN

BORK ROOMED IN A SHABBY HOUSE IN A SHABbier street, as the only lodger of a deaf and half blind old man, by the name of Wilson, who saw him only once a month to collect the rent. Old Wilson seldom knew when his lodger entered or left the house, and he cared less. What the old man had done for a living in his prime was more or less of a mystery. Report had it that he was a broken down Alaskan miner who had made and lost a dozen fortunes. Before going completely broke he had bought himself a shack of a house and invested his remaining capital in government bonds, on the meagre income of which, and the rent from the upstairs spare room, he eked out a Spartan existence. The place was ideal for Bork, who hated the habitual prying of even the most reserved landladies. Old Wilson never entered his lodger's room. Consequently it was cluttered with empty bottles shamelessly exposed in the most conspicuous spots.

In spite of his age, Wilson was not an early riser. He enjoyed his ten hours in bed best of the twenty-four. Doubtless the futility of being up and about, when he could see but little and hear less, impressed on him the wisdom of dozing away as much of his meaningless life as possible.

About half past four on the morning of his mad escapade in the laboratory, Bork stumbled up the rickety stairs to his room. The necessity for an alibi in case of investigations regarding the shattered tube was beginning to dawn on him. He knew that old Wilson would not hear him, so he made no effort to walk softly. The alibi presented itself ready made. At seven o'clock Wilson's customary hour for rising, Bork would hunt up the old man in his kitchen and pay the rent a day in advance. Then the old fellow could swear that Bork had spent the night in his room, and believe his oath. The early payment of the rent would arouse no suspicion, as Bork had frequently paid a day or two ahead of time.

Opening the door of his room, Bork found the light still on. A half quart of whiskey stood on the untidy bureau. It was but natural in his shattered state that he should take a bracer to steady his lacerated nerves. He poured himself a stiff jolt and raised it to his lips. As the reek of the crude alcohol fumed his nostrils he was overcome by a strong feeling of revulsion. Yet he imagined that he needed the drink desperately. His attempt to swallow it proved unsuccessful; his body simply rejected the proffered mercy. A healthy young savage almost invariably rebels against his first swallow of raw whiskey, whatever may be his reactions to his hundredth. Bork was in precisely the same condition, except that his aversion was a thousandfold more intense. To down a drink in his present condition was impossible. Instinctively he hurled tumbler and bottle to the floor, smashing three empties in the act.

"What a fool I've been," he muttered. "I must have been sick as a dog to like that stuff."

Aware of an indefinable sense of power, he clenched and unclenched his fists, watching the ripple of the firm muscles beneath the skin. Presently he started. His hands, ordinarily a pasty yellow, were tanned a deep, healthy brown. He might have been working for months outdoors beneath a tropical sun.

A startled glance in the shaving mirror above the bureau confirmed his half formed suspicion. His face was as swarthy as a Hindoo's, and his yellow, fine hair had turned jet black and as coarse as an Indian's.

Even these radical changes, however, failed to account for the utter difference between the face staring wide-eyed from the dusty mirror and the familiar features which he remembered as his own. A more fundamental alteration had transformed his appearance completely. Suddenly he recognized its nature. His blue, cold eyes had turned black and strangely luminous. With a terrific shock he perceived also that he appeared to have grown younger.

In silence he slowly began removing his clothes. Five hours before his body had been like a young boy's, smooth, white, and practically hairless. Now his skin was the same rich brown hue, from heels to head, as his face and hands. Moreover, his chest, arms and legs were covered by a thick growth of coarse black hair like a professional weight-lifter's. From skin to marrow he was physically a different man. No one who had known him intimately five hours previously could have identified him as Neils Bork. This was a different man.

"Who am I?" he asked aloud, reaching for his shirt.

No sooner were the words out of his mouth than he realized two further changes from the man he had been, each of the profoundest significance. The querulous voice of Neils Bork had deepened and become vibrantly resonant. It was the voice of a man with both strength and personality and an assured confidence in his power to use them to his own advantage. It also was a voice that would attract women. Second, he noticed a new trick of habit, that was to become instinctive. The hand reaching for the shirt drew back, and for a simple, natural reason. The shirt was soiled. To put such a thing next to his skin—Bork had never worn underclothes—was an impossibility to the new man.

From the bureau drawer he selected his best shirt, a white

linen freshly laundered, which he had worn but once or twice. The shabby suit was discarded in favor of his single decent one, a gray tropical wool. This he had not worn for over a year. The cut was a trifle out of date. That, however was of no consequence. The suit was wearable, having been dry-cleaned before it was put away. Clean socks, his best shoes and a plain black scarf, which he had discarded as being too tame after one wearing, completed his outfit. It did not occur to him to seek a hat; his thick, black hair afforded ample protection from sun and weather.

Although the old Bork had been a heavy, consistent drinker, he was not absolutely thriftless. From a slovenly suitcase stuffed with soiled clothes and worthless letters from girls a little less than worthless, the new man extracted the "dead" Bork's carefully hoarded savings. These amounted to about six hundred dollars in ten and twenty-dollar bills.

By a quarter past five the new man was ready to face his dawning life. Thrusting the roll of bills into a trousers pocket, he started for the door. Then he remembered poor old Wilson's rent. Although neither alibi nor disguise was now necessary, the new man felt that it would be wise to dispose of the old forever. Having found a stub of pencil, he scribbled a note on the back of some forgotten girl's scented envelope.

> *"Mr. Wilson: The enclosed ten dollars is for the month's rent I owe you. As you will see from my room, I have been a steady drinker. The stuff has got me at last. Rather than give my boss the satisfaction of firing me, as he must sooner or later, I am firing myself. Give this note to the police. They will find my body in the Pacific Ocean if they want it, and if the crabs don't get it first.*
>
> *Neils Bork."*

On his way out of the house he slipped the envelope with the ten-dollar bill under the kitchen door. Old Wilson was not yet stirring. The new man took with him nothing but his money and the clothes he wore. No one saw him leave the house; it was still too early for decent workers to be going about their business. He strode briskly along in the clean morning air, conscious of a new vitality coursing through his veins like the elixir of life itself.

While the man who had been Bork was confidently marching to meet his destiny, Crane lay tossing and muttering in his fitful sleep, tormented again by the prickling of his skin. Shortly after six o'clock he awoke fully and leapt from his bed. The itching was much less severe than the first attack. Nevertheless it was sufficiently distressing to make him hurry his dressing and rush to the doctor's house.

Brown had not gone to bed. The curious glance he shot Crane was almost hostile.

"What's up?" the latter enquired, feeling the doctor's restraint.

"That's what I want to know," Brown answered shortly. "Where have you been the past week?"

"At my usual stand," Crane grinned. "The workshop of the Foundation, the twenty million volt laboratory, at home in bed, and up the street three times a day on an average for my meals."

"Is that all?" the doctor demanded suspiciously.

"Sure. Where did you think I'd been?"

"I couldn't guess," the doctor replied slowly, "unless it might have been some low dive of Mexicans or Orientals. Whatever it may be that you've got yourself infected with is new to any science I know. Your case is unique. Itching again?"

Crane nodded. "Save the lecture till after you've cured me. Then I'll listen and admit anything you like."

"The cure will be easy enough. You must soak yourself in disinfectants till the last particle of scale or dust is sterilized and removed from your skin. It may be a long job. Take boiling hot baths and make yourself perspire freely before you rub down with the disinfectants. Then do it all over again two hours later. Keep at it until the itching stops completely. I'll write out the prescriptions."

"You seem sore about something," Crane remarked as the doctor handed him a sheaf of prescriptions. "Why don't you speak up and get it off your chest?"

"I will when I know what it is myself. You give me your word that you are telling the truth about what you have been doing?"

"Of course. Why should I lie? If you want to check up, ask the men at the Foundation and my landlady."

"Then," said Brown, "we are going to discover something brand new. By the way, I should like to examine your assistant —Bork, isn't it? Has he been working with you all the time?"

"In the shops, yes. But not in the twenty million volt laboratory. I must have put in thirty hours with my two million volt tube going at full blast during the past eight months. Bork hasn't been around it more than an hour all told at the most."

Brown considered in silence for some moments.

"Will you let me try an experiment with your tube?"

"Any time you like, if I handle the switches. I would let you do it yourself if you had worked around high tension apparatus. You give the instructions and I'll deliver what you want."

"Very well. How about one o'clock this afternoon? I must get some sleep first. You can go home and begin sterilizing your skin."

"That will suit me. I'll be in the high tension laboratory at one o'clock. Ring the bell and I will let you in. You can't give me a hint of what you think you're doing?"

"I could. But why go off half-cocked? Either this is the biggest thing in a thousand years, or I'm completely fooled. This afternoon will decide."

Doctor Brown's discovery was not, however, to receive its test so soon. The late Neils Bork had made that impossible. Many weary months of toil and speculation were to pass before Brown could recapture the first fine careless rapture of his glimpse at a new universe.

The doctor had intended going straight to bed the moment Crane left. Sleep, he now realized, was out of the question. To quiet his busy mind he must pacify it by consulting the voluminous biological literature that would at least eliminate the chance of making crude, ignorant blunders.

By good fortune the Aesculapian Society had a prosperous branch in Seattle with an excellent scientific library covering all phases of biology and medicine. Doctor Brown was local vice-president of the Society. His pass-key would admit him to the library at any hour of the day or night, a privilege not shared by ordinary members. He shaved, asked the housekeeper for a cup of coffee, pocketed the rough sketches of what he had seen under the microscope, and hurried off to the biological library. There he spent five hours from seven till noon poring over biological atlases and massive treatises on the protozoa—those simplest of all animals.

With a sigh, as the clock chimed out a musical twelve, he closed his books and rose to prepare for his appointment with Crane. All his painstaking search had so far yielded no glance of a similarity between what was already known to science and what he had discovered. Thus far the outcome was encouraging. But the mass of ascertained fact about the humblest living creatures is so enormous that Brown estimated his good fortune at its precise value—nothing. A search of weeks would be necessary before he could assert confidently that he had made a vital discovery—if indeed he had.

His preparations for the approaching experiment with Crane were simple in the extreme. Having lunched at a cafeteria, he asked one of the girls behind the counter for a dozen raw, new laid eggs. Half of these were for the test, the other six to control the test. Part of the necessary apparatus he hoped to find in Crane's laboratory; the rest he must provide himself.

"Where can I buy a hen?" he asked the girl.

Thinking him slightly mad, the girl replied that there was a poultry market six blocks up the street. Brown thanked her and drove to the market. There he purchased the most motherly looking, clucking Buff Orpington on exhibit, a large slat coop to house her in, and loaded her on the back seat of his open car.

The spectacle of the well-known Doctor Brown threading his way through the traffic with an eloquent brown hen as passenger caused several traffic jams. However, he got his collaborator home safely and turned her loose in the walled back garden. Before leaving her to enjoy the tender young zinnia seedlings, he made a passable nest of excelsior in the slat cage and presented the prospective mother with half a dozen new laid eggs. With that attention to details which is half of scientific success, the doctor marked an indelible blue cross on each of the eggs, so that the 'controls' should not be lost among the mother's possible contributions.

"Do your stuff, Bertha," he counselled, carefully disposing three of the remaining half dozen eggs in each side pocket, "and I'll do mine. Good bye; I'm half an hour late already."

During the five hours that Brown was winding his devious way through mazes of the protozoa in the Aesculapian library, the man who had been Bork made rapid explorations into the wonders and mysteries of his transmuted personality. On reaching the main business street nearest his former lodging, he eagerly sought out a restaurant. The old Bork had always fought shy of breakfast, for obvious reasons. The new man was ravenously hungry. It was still very early. In his rapid walk he passed several cheap, all-night lunch counters, hesitated for a moment before each, and quickened his pace, to leave them behind as rapidly as possible. This swarthy young man with the strangely luminous eyes was fastidious to a fault.

At last he found what he wanted, a spotlessly clean, airy lunch room with white glass tables and a long cooking range under a hood and in full view of the customers. A girl in a white cap and clean white smock, her arms bare to the shoulders, was deftly turning flapjacks on a gasplate by the window. As the new customer entered, she glanced up from her work. Ordinarily a second's inspection of the men who passed her by the hundred every day satisfied her curiosity. There was an undefinable 'something' about this new man, however, which riveted her attention instantly. Unconscious that he was being watched, the swarthy young man walked to the far end of the room and sat down at a small table. A smell of burning hot cakes brought the girl out of her dream.

"That's somebody," she remarked to herself, but half aware of what she meant. She was right. This man was 'somebody,' not a mere 'anybody' undistinguishable in any significant way from tens of millions as commonplace as himself.

The 'somebody' was giving his order to an elderly man waiter who stood, pad in hand, trying to concentrate on his job. But he could not.

"Excuse me, sir," the waiter began diffidently, "but haven't I seen you in the pictures?"

"Pictures? I'm afraid I don't understand."

"The movies, I meant."

The 'somebody' threw back his head and roared with a deep, resonant laughter. It was an echo of the laughter of the gods. Early breakfasters turned in their chairs fascinated and amused by that hearty, good-natured shout, wondering what the joke was. Then they saw the young man's face and studied it openly, curiously. What was there about him that instantly attracted all who got a square look at him?

The embarrassed waiter stammered an apology.

"That's all right. No, you have never seen me in the movies, and I hope nobody ever will. Ask the cook to hurry that order, like a good fellow, will you? I'm starving."

Still unconscious of the sensation he was causing, the stranger casually inspected the simple decorations and general arrangement of the room.

"It just misses being good," he thought. "What's wrong? Everything is clean enough, and the fresco doesn't jar like most."

The day manager had just arrived and was taking up his position behind the glass cigar counter. Seeing a new customer, and a distinguished-looking one, apparently in need of attention, he walked down to inquire what he wished.

"Have you been waited on?"

"Yes, thanks."

"Is there anything I could do for you?"

"Probably not. It's too late now. I was just thinking that the architect missed a masterpiece by a mere hair-breadth. If this room were three feet longer, two feet narrower, and seven inchs higher, the proportions would be perfect. The tables don't fit, even now," he went on, unconscious of the astonishment on the manager's face. "Don't you see," he continued earnestly, "what a stupid waste it is to bungle a room that might be perfect if only a little forethought—not afterthought—were given to its design?"

"Excuse me, but are you an artist?"

Again the answer was a shout of laughter.

"No," the swarthy young man replied, subsiding. "And I never will be so long as I keep what mind I have."

"Then may I ask what your profession is, Mr.——?"

The luminous black eyes seemed for a second to look back and inward. What name should he give himself? For the moment, although his mind worked at lightning speed, the new man had difficulty in recalling what his name had been.

Bork; that was it. Obviously it must be discarded.

His eyes roved to the cigar counter.

"De Soto," he said, slightly altering the name of a popular cigar advertised in red and gold on a placard behind the counter. The name would fit Portugal, Mexico, Spain or South America and definitely rule out the Orient. Some touch of the torrid south was necessary to explain his coloring. An Oriental in America might reasonably expect hostility in certain States, while Latins or South Americans would be accepted as human beings.

It was a wise choice.

"I hope you will drop in often, Mr. De Soto," the manager replied with more than the perfunctory courtesy of business. He meant exactly what he said, for he, too, felt a subtle attraction to this dark young fellow who was the picture of health, and who could laugh so infectiously.

"I shall be delighted, whenever I am in the neighborhood."

His breakfast arriving at that moment, further exchange of civilities was left to the future, and the manager retired to inspect the ledger.

" 'De Soto,' " he muttered to himself; "where have I heard that name before? Wasn't there an early Spanish explorer of that name? Still, this boy doesn't look Spanish, and he hasn't the trace of a foreign accent. He's somebody, whoever he is. I must ask him when he goes out what he does."

De Soto's breakfast was perfectly cooked and beautifully served, from the rare tenderloin steak and crisp French fried potatoes to the sliced oranges. It was the breakfast that a strong laboring man would have ordered—if he could have afforded to pay for it. No modern business man or sedentary scholar could have looked it in the face. De Soto disposed of the last morsel. There remained only the black coffee—a reminiscence of the dead, shaky Bork. At the first sip De Soto hastily set down the cup. Even this mild stimulant reacted instantly on his perfectly tuned body. It was impossible for him to touch the stuff, and he finished his breakfast by drinking three glasses of water.

The sip of coffee had a curious effect on De Soto. Of what did it remind him? Someone he had known? No; that wasn't it. Everyone at the tables near his was drinking coffee and apparently enjoying it. Evidently it was a common and harmless indulgence. Then what was it that he was struggling to recall? It concerned himself, personally and intimately. Of that he felt certain. From the fast receding life which he had left behind him forever, a voice like that of a drowning man whispered "Bork". De Soto half remembered in a strangely inaccurate fashion.

" 'Bork'?" he muttered to himself. "Who was Bork? Ah, I begin to remember. He was the man—electrician, or something of that sort—who committed sucide some months ago by drowning himself. Where did he do it? I seem to recall that he drowned himself in his room, but that's impossible. I've got it! The Pacific Ocean. Was it that? How could it be—it's too vague. The Pacific Ocean might mean anything from here to—."

His thoughts broke off abruptly, baffled by his inability to recall "China". Not only his own past was being rapidly swallowed up in a devouring blank, but also much of his elementary knowledge of the world which he had acquired as a schoolboy. For a moment he felt mentally ill. He knew that he should remember, and wondered why he could not. Some silent comforter put solace in his way.

"I cannot have lost all my mental habits," he thought. Setting his teeth, he reached for the menu. "Can I still read? If not, my intellect is gone."

He opened the elaborate card. The effect was electric. Instantly the long pages of close print registered on his mind as on a photographic plate. Without the slightest conscious effort he had read and memorized two large pages of heterogeneous, disconnected items in a single glance. He smiled and reached for the morning paper which the man at the adjoining table had left behind. As fast as he could turn the thirty-two crowded pages he scanned them at a glance, photographed every item, whether news or advertisements, indelibly on his consciousness, and digested the meaning of all. Curiously enough he believed that he had "always" read in this manner. An apparent inconsistency, however, caused him a moment's uneasiness. There was much in the morning's news about the fighting in China. Reading of China, he visualized instantly all that he had ever known or imagined about that country and its people. Why then was he unable to remember the name when he tried consciously to recall it?

"I must have the stimulus of innumerable associations to think about any one thing, I suppose," he mused. "What was I going to do when I came in here? Breakfast, of course. But what had I planned to do next? It was connected with that man Bork's trade. Electrician. That was it. I know now; I was on my way to ask for work where I can study electricity, X-rays, and all that sort of thing. Why, I have always dabbled in electricity. How stupid of me to forget. My stomach must be badly upset. Well, I'll dabble no longer. This time I go into it for all I'm worth. Where the deuce was I going?"

Suddenly the name Crane flashed into his mind. For no reason that he was capable of discovering, De Soto began to rock with uncontrollable laughter. There was a tremendous joke somewhere, but what it was all about he could not for the life of him say. Nevertheless he continued to shout with jovial laughter till the whole restaurant turned to stare at him. Aware of their puzzled faces, he made a pretense of reading the comic strips of the paper and controlled himself. It was time to escape before some shrewd busybody should guess the secret of his joke —which he did not know himself. That was the funniest part of it. Who and what was Crane? Whatever the elusive Crane might be, he was at the bottom of De Soto's haunting, mysterious joke. Calming himself, he beckoned to the waiter.

"Will you pay the cashier? I have no small change. Please

bring me a telephone directory."

When the waiter returned with the change and the directory, De Soto tipped him generously and proceeded to look up Crane. There were several Cranes listed. Their first names or initials all seemed somehow wrong. De Soto closed the directory and let his mind drift. A single glance had sufficed to print the entire list of Cranes, their addresses and occupations, on the sensitive retina of his mind. One name, Andrew Crane, room 209, Erickson Foundation, seemed to stand out from all the others. Why? What was the Erickson Foundation? He decided to ask the manager on his way out.

"Can you tell me where the Erickson Foundation is?"

The manager gave clear directions for reaching it.

"What sort of a place is it?" De Soto asked. His tone implied that he wished to learn the public estimate of the Foundation, not what its specialty was. The latter, De Soto felt, might be a suspiciously ignorant question. Some monitor was prompting him to use caution; why, he could not fathom.

The manager, bursting with civic pride, enlarged upon the world fame of the Foundation, which was heavily advertised in the local papers. He even boasted that Doctor Crane had made the most powerful X-ray tube in the world, and was now nearly ready with a giant that would surpass anything the Foundation's jealous rivals could hope to produce for a hundred years—perhaps for two hundred. De Soto struggled again with that awkward impulse to burst out laughing. The manager concluded his booster talk by a direct personal question.

"Are you in the electrical line?"

"Only a student," De Soto replied instantly. "I plan to go into X-ray work as soon as I have finished my course." This straightforward reply seemed, to the man who gave it, to be the simple statement of an ambition which he had "always" held.

"Where are you studying, if I may ask?"

De Soto hesitated, nonplussed. Where, exactly, was he studying? Had he ever studied?

"Oh, I'm just reading by myself."

The moment he had uttered the words, De Soto knew that he had told a falsehood. Instantly he corrected himself. It seemed the only natural thing to do; the lie tasted worse than coffee.

"I meant to say," he apologized, "that I'm going to start my reading this morning."

"Oh," said the manager. Then, irresistibly attracted by this frank young man with the singularly penetrating black eyes, he added a word of heartfelt encouragement. "You'll make good. Some day we'll hear of you in the Foundation. Drop in again."

De Soto left the restaurant, followed by the hungry eyes of the flapjack girl and by those of every other woman in the place. Although he was now so far ahead of his past that to look back on Bork was impossible, he had a strange feeling of "difference". From whom was he different? The faces of the men and women he met gave him no clue. Many of the easy going Mexicans had hair as black as his, skins even darker, and eyes almost as black. Could he have remembered Bork's features to compare them with his own, reflected in the shop windows he passed, he would have noticed that his lips were fuller and redder than Bork's, and his nose more thoroughbred-looking than Bork's had ever been. His own nostrils were slightly distended, like those of a race horse eager to fill its lungs with all the fresh air blowing in its face, not close and pinched like Bork's. The whole "set" of the face bore not the slightest resemblance to that of the "dead" man. This was a countenance alive with purpose and the will to achieve it; the other was the peevish mask of a neurotic weakling. No feature of the "dead" man survived as it had been, and no trick of expression remained to betray the future to the past. But of this De Soto was wholly unconscious.

"Can you direct me to the Public Library, please?" he asked a traffic officer.

"Two blocks north—that way; three west."

He reached the library just as the doors were opened. At the reference desk he asked where the books on electricity and X-rays were kept. To his surprise the middle-aged woman in charge did not reply immediately. She could not. Something in this strange young man's face reminded her of a boy who had been dead twenty-five years. There was not the slightest physical resemblance between the features of this dark-skinned, singularly intelligent-looking young man and those of the boy whose face had almost faded from her memory. Yet the one passion of her desolate life flamed up at her again as if it had never died.

"I'll show you where they are, Mr.——"

"De Soto," he supplied, wondering why she fished for his name.

"Pardon me for asking, but are you related to the Stanley Wilshires?"

"Not that I know of. I was born in Buenos Aires, and all my people have lived there for generations. Why do you ask?"

"You reminded me of someone—not your face, but your look. I must have been mistaken. Here are the electrical books. The X-ray material is in the next stack. You may use that table if you wish."

"Thank you. I shall probably stay here all day, as I have a lot of reading to do."

The librarian walked thoughtfully back to her desk. "I could have sworn it was Frank looking at me again. How silly!"

Many a woman was to experience a like feeling on first seeing De Soto's face. What attracted them, or what recalled the men they had loved, they could not have said, for it was beyond simple analysis. It was not an instinct for sympathy, for De Soto's face suggested confident strength rather than sympathy. Possibly the secret lay in the message of superb, clear vitality that shone from his eyes, recalling the heightened manhood with which these starvelings, in the fondness of their imagination, had once endowed their lovers—before they learned that fact and wishful fancy are a universe apart. Here was a man who lived with his whole body, as they would have wished their lovers to live, not rotting by inches year after year into a lump that was human only in name.

While De Soto in the Public Library, and Doctor Brown in the Aesculapian, unaware of one another's existence, were unconsciously storing up ammunition for a grand assault on one of time's deepest mysteries, old Wilson precipitated the official suicide of Bork. On rising at seven o'clock to prepare his breakfast, the old fellow found the envelope with the rent. His eyes were still good enough to make out, in a hazy way, a ten-dollar bill. They also perceived that there was writing on the envelope. Poor old Wilson had a premonition of the truth: his precious lodger had decamped, and this was his heartless way of breaking the terrible news. Without stopping to get his breakfast he doddered over to his nearest neighbor's.

The obliging neighbor shouted the dire message, a word at a time, into Wilson's better ear.

"You had better tell the police at once."

"Eh?"

"The police. Tell them at once."

"You do it. I've got to find another fool to rent my room."

The neighbor obliged the old man willingly. The service would get his own name into print, a distinction which he had not yet enjoyed.

The ten o'clock "noon edition" of the evening papers gave the sublimated Bork a generous headline and toyed in audible whispers with the dead man's shocking allusion to crabs. They also, one and all, lamented the regrettable and undeserved notoriety which this suicide would bring upon the world-famous

Erickson Foundation. For they had grown just a little tired of tooting the Foundation's siren for nothing, and decided now to recover their just dues by spreading some real news. To be worth printing, newspaper personalia must be spiced with more than a hint of scandal.

The Director of the Foundation first learned the facts from one of his enemies with a low taste for extras. This gentleman telephoned his sympathy—"I have just heard the distressing news. Oh, don't you know? One of your staff has committed suicide in a shocking manner—gave himself to the crabs. Who was it? Neils Bork. You say he was only a technical assistant? The newspapers don't mention that. This will be a terrible blow to the good name of the Foundation. Well, you may count on me to do what I can."

The Director preferred not to count on his friends. Instead, he got Crane on the telephone—following a long wait during which Crane hastily dried himself after his second stewing.

"What's all this about Bork committing suicide?" the Director snapped.

"Suicide? I know nothing about it. You must be mistaken.

"The papers are full of it."

"Excuse me a minute. This is a shock."

Crane sat down suddenly on the chair beside the telephone and buried his face in his hands. In spite of his "kidding", he had liked Bork. For perhaps the first time he realized fully how deeply attached he had been to the surly fellow whom he had tried his best to make something of. Although reason convinced him that bad whiskey and not he was responsible for the tragedy, a deeper voice accused him relentlessly.

"If it is true, I'll see what can be done."

"Very well. Please come to my office as soon as you can. This will give us a black eye with the trustees unless we can hush it up."

"I'll be down in half an hour."

4

THE WIDOW'S REVENGE

ON REACHING THE LABORATORY SHORTLY after one-thirty, Doctor Brown found a note addressed to himself stuck to the steel door by a strip of adhesive tape. It was from Crane, stating that he was "in conference" with the director, Mr. Kent, and asking the doctor to come to Kent's private office.

Not having seen the extras featuring Bork's suicide, Brown wondered what was up. "In conference" usually meant a wigging for some unfortunate member of the staff. Brown knew Kent well and did not exactly respect him. Kent, whose talents were purely political and administrative, boasted only a high school education. For his particular job he was competent enough, although his outlook was essentially unscientific. In matters of unimportant detail he was a martinet of an extremely exasperating type. Being what he was, and not being what he wasn't, Kent found his domineering fussiness and his social cowardice hotly resented by the eighty highly trained men under his alleged control. But, as he was the chosen of the trustees, the scientists of the Foundation had to grin and bear him. In all fairness to Kent it must be admitted that he was highly efficient in the particular task for which the trustees had picked him. This was the rather ticklish job of coaxing superfluous cash from retired millionaires, who nervously foresaw their rapidly approaching passage through the needle's eye mentioned in Scripture.

The eighty under Kent's nominal kingship treated their malevolent despot with a mixture of amusement and contempt. As the least of them had three or four times the ingenuity and imagination that Kent could claim, they made of his life a very creditable imitation of hell. Baiting Kent became the favorite pastime of their idler moments. When research palled, these misguided men would put one of their number up to making some perfectly outrageous demand of the harassed director. Then, when Kent naturally refused, the petitioner would indignantly "resign". This always brought Kent to his knees instantly. To go before the trustees and admit that the Erickson Foundation was not a cooing dovecote of high-minded scientists, who were toiling only for the good of humanity under their director's brooding benevolence, was more than the poor man could face. Harmony and cooperation, service and uplifting self-sacrifice being the director's official slogan, he dared not confess that "his" men were a thoroughly human lot, with all the self-interest of the average man, and a perfect genius for making themselves, on occasion, as irritating as a pack of discontented devils. Kent would have been happier as manager of a five-and-ten-cent store, where he could have hired and fired with Jovian irresponsibility.

In spite of his tactlessness and his aggressive stupidity, Kent had one feature—if it can be called that—which redeemed him almost completely in the eyes of his subordinates. His nineteen-year-old daughter Alice, fair-haired, violet-eyed and altogether wholesome with her keen sense of humor, was adored by every man on the staff, from the tottering De Vries, seventy years old, but still with the mind of a man of forty, to the gangling youths fresh from the university and just emerging from their rah-rah stage of development.

But for the alluring Alice, it is doubtful whether a single

member of the staff would ever have attended the tiresome teas and deadly dinners which Kent imagined it his official duty to inflict on his imagined slaves. Even a funeral or a college commencement, they agreed, would have been lively with Alice as hostess. Kent, for his part, worshiped her from heels to hair. It was rather pathetic to see the jealous care with which he hovered about her when some attractive young man seemed to be getting on too fast in her affections. The husband of Alice, if poor, futile Kent could choose him for her, would be so impossibly perfect as to be a mere platonic ideal. He secretly hoped that her disconcerting sense of humor would keep her a spinstress, at least until he was a handful of ashes in a white jar.

On entering the holy of holies, Brown found Kent and Crane glowering at each other across the broad expanse of a mahogany table that looked almost as if it were not veneer. Sensing that the interview was now at the resignation point so far as it concerned Crane, the doctor made a motion to withdraw, but Crane irritably motioned him to a chair.

"There's nothing private," he announced. "Mr. Kent and I have been discussing Bork's suicide. I'll be ready for your experiment in a moment—if Mr. Kent doesn't force me to resign."

"Bork's suicide?" the doctor echoed. Brown seldom bought an extra; the war had cured him of the vice.

"Yes. Last night. The facts seem to be clear. It's the ethics of the situation that are worrying Mr. Kent and me. I see it one way; he, another."

"You might compromise," Brown suggested. It was not the first time he had acted as liaison officer between Kent and his unruly staff.

"Precisely," Kent took him up eagerly. "We must cooperate. Doctor Crane, unfortunately, refuses to see the absolute justice of my stand."

"If he did," Brown smiled, "he would have to surrender, wouldn't he? I shouldn't call that much of a compromise. It takes two to dispose of cold crow, you know—one to dress it, the other to eat it. Who is the chef in this instance?"

Crane's long jaw set obstinately.

"I am," he assented defiantly. "Mr. Kent insists that the good name of the Foundation be preserved at all costs—even that of common decency to a dead man, who, naturally, can't speak in his own defense."

"Excuse me a moment," Brown interrupted, "but aren't you beginning to itch again? You see," he continued, turning to the glowering Kent, "Doctor Crane is rather irritable today. By working around his two million volt tube without even the precautions of common sense, he has got his skin and his temper into a very ticklish condition. Nothing serious, of course; merely hard on the disposition. So you will pardon his rudeness," the doctor concluded with a disarming smile, "if he forgets himself. Very well, Doctor Crane, go on; sorry I interrupted you."

"Mr. Kent has been listening to a lot of old wives' gossip. Somehow it had got around that Bork was a soak. Can you prove it, Mr. Kent? No? Well, then shut up. I mean exactly what I say," Crane continued, lashing himself into a passion and entirely disregarding the doctor's warning look. "If you dare to give the papers any such scandalous lie about Bork, I'll give them a better one about how you run this Foundation. You're not going to throw mud all over a dead man's name just to save what you think is the honor—it hasn't any—of this corporation. All the Erickson Foundation gives a damn about is the patents it gets out of its employees for the usual nominal fee of one dollar—and you know it. There is no question of ethics here. You have none, the Foundation hasn't any, and I don't know what the word means. Get the point? The best man—or the biggest crook—wins. That's all. This time I win; you lose. You give it out to the papers that Bork was temporarily insane from a nervous breakdown, or I'll resign and tell the yellow rags why."

Crane had gone too far. His itching skin had betrayed him into a complete statement of the contempt—justified, perhaps—in which he held his job, the director, and the canny corporation for which he worked. The fact that he obviously meant every word he said did not lessen the enormity of his offense in the eyes of the outraged director.

"This," said Kent in a cold rage, "is insubordination. I would be quite justified in recommending your instant dismissal to the trustees."

Crane gave a short, contemptuous laugh.

"What do I care for your silly job? I can get a better one tomorrow. You know as well as I do that I'm the best X-ray man in the country. You see, I'm not fainting from false modesty. Call in your stenographer and dictate that letter to the papers. Otherwise you can have my resignation here and now. That's final."

"You heard him," Kent exploded, turning to the embarrassed doctor. "Is that a proper way for a subordinate to address his superior? I shall report him to the trustees. You are my witness."

" 'Superior' be damned," Crane cut in before Brown could attempt to restore diplomatic relations. "Are you going to dictate that letter to the press? Yes or no?"

"No," Kent snapped. But it was a half-hearted snap, such as an aged and ailing turtle might have given his persecutor.

"Then I resign," Crane announced, rising to his feet. "Explain why to your precious trustees. If you don't, I will."

"Sit down," the doctor commanded sharply. "You are both making fools of yourselves. The only excuse for you, Crane, is that you have let a little itching get the better of your temper. I know personally," the doctor continued, turning to the enraged director, "that Doctor Crane has the very highest regard for you personally and for your amazing success in running the Foundation. Can you afford to lose such a man—your most loyal collaborator? You know you can't. Why not compromise? Crane agrees to stay in exchange for a short statement from you to the press clearing Bork's name. Bork, I gather, has committed suicide in a fit of insanity brought on by overwork. Why not state the simple fact plainly, Mr. Kent? It is no reflection on the Foundation or on your policies. Hundreds of men work themselves to death every year in the United States alone. And why not? It's better than ossifying."

Kent glared at Crane, a hard, newborn hatred such as he had never before experienced toward his "subordinate" wrestling with his common sense. The thought that Alice seemed to prefer this lank, outspoken Texan with the uncompromising jaw to any of the other younger men on the staff, but added more fuel to Kent's cold, smouldering rage. By openly defying his nominal superior, and showing him up in the presence of a third party for the overstuffed effigy which he was, Crane had earned the director's lifelong enmity. Kent was shrewd enough in a shoddy, political way in dealing with human nature. He resolved on the spot to do everything in his petty power, by underhand suggestion and faint praise, to turn his daughter against this man for whom she had more than an incipient fancy. A dozen promising attacks flashed across his narrow mind as he scanned Crane's frankly scowling face, his own gradually assuming the bland benevolence of a well-steamed suet pudding. In more senses than one it was a historic moment, although neither man could possibly have foreseen the strange consequence of their mutual folly. Doctor Brown, mistakenly inferring from the director's expressionless face that the storm was over, managed to wink at Crane unobserved by Kent.

"I withdraw my resignation," Crane mumbled, rightly interpreting the doctor's wink. This row was to end as all such rows invariably ended, in a climbdown by the director. So Crane in his easy self-confidence imagined. As a matter of fact, he had not only cooked his own goose to a turn, but the unsuspecting Alice's as well.

Kent apparently swallowed the bait—again as usual. The magnanimous director rose from his swivel chair and walked clear round the table to extend to the contrite Crane the manly hand of forgiveness.

"And I," he promised with pompous solemnity, "will give the papers the simple truth that Neils Bork died a martyr to science, betrayed to death by his own zeal in the service of knowledge and the pursuit of truth." The director inflated his chest, and Crane smothered a grin. "As you have well said, Doctor Brown," he continued proudly, "the shoulders of the Erickson Foundation are broad enough, and strong enough, to support the truth, the whole truth, and nothing but the truth."

It was a rotund utterance, worthy of the fattest of the innumerable commencement addresses which Kent was in the habit of inflicting (by invitation of boards of trustees as progressive as his own) on successive generations of young skeptics, all over the United States, who have outgrown this particular brand of twaddle. Yet, such an indurated ass was the good director that for the holy moment he believed every word he said. What he truly meant, in his subconscious mind, was roughly as follows: "You, young man, have made a fool of me before this doctor friend of yours. I shall wait until your back is turned. Then I will stick my longest knife clear through you. And it will hurt, for I shall turn it."

Feeling that the honorable director was, after all rather a slippery mackerel, the conspirators deferred their departure until Crane had the meticulously dictated lie to the press firmly in his competent right hand. As a further proof of their sound appreciation of the good faith of the born administrator, they lingered until the message was safely delivered to an eager reporter summoned by telephone. Then, and then only, did they venture to take their respectful leave.

"And to think," Crane remarked viciously, as the massive door of the inner sanctum closed noiselessly behind them, "that he is the father of Alice. It's impossible."

"Heredity is a theory," Brown admitted, "in this case, any way, and environment an illusion. The Kent family alone would disprove half of our biological guesses. And I'm going to take a crack at the other half."

To Crane's wondering eyes the doctor complacently exhibited six large, clean hen's eggs.

"Here's my apparatus. You have heard of Watson's experiments with fruit flies? Well, I'm going to try something similar. Flies don't prove very much for human beings. Hens are nearer our own kind."

"But a hen is essentially a reptile," Crane objected vaguely, as some shadowy reminiscence of his high school biology flitted across his electrical mind. "Birds came from snakes, didn't they? And we branched off from monkeys."

"Go far enough back," Brown suggested lightly, "and you'll find our common ancestor where the mammals sprang from the effete reptiles. No matter how far back you look—provided you stop within a few ages of the protozoa, you won't find any of our cousins among the insects. We may not resemble hens very closely, but we are more like fowls than we are like flies. Watson's work was great as a beginning. It was tremendous. He discovered a new world. I want to take the next obvious step. Is your tube working all right?"

"It was when we quit yesterday," Crane replied, inserting his key in the lock of the green steel door. "What an awful smell! Do you get it?"

Brown sniffed critically.

"That's organic matter decaying. I should say——."

What he was about to add remained unsaid. A terrific odor gushed out at them, beating back their feeble assault upon the pitch darkness of the laboratory. Not to be defeated by a mere smell, Crane flung a flap of his coat over his nose and mouth, and groped desperately for the switch controlling the floodlights. The switch clicked its futile message.

"Dead," Crane exclaimed, referring to the lights.

"Check," Brown muttered through his handkerchief, imagining that Crane referred to the overpowering reek. "This is more than organic decay. Don't you get a metallic taste as well? What on earth——"

Brown's further observations were strangled in an involuntary croak of instinctive horror. He had followed Crane along the palpitating darkness of the steel gallery of the switches when suddenly, from the sheer blackness above him, eight clammy "arms", colder than the black death and slimier than a decaying tangle of kelps, descended upon his head, chest and shoulders in a loathsome embrace. Instantly he was out of the laboratory, struggling like a madman to free himself from that frigid abomination adhering to the upper part of his body with all the chill tenacity of a dead octopus. Crane followed, shaking from head to foot.

Once in the glaring sunlight, the horrified men saw in a flash what had descended upon the doctor. Crane tugged it from the doctor's shoulders and hurled it away with a shudder of disgust. It was a hideous knot of eight smooth, slimy black legs, each about four feet long, still adhering crazily to the tattered fragment of a huge black thorax on which the dull red imprint of an "hour glass" was plainly visible. The rest of the monster's body had already evaporated in foul decay.

In petrified fear they stood staring at the remains of this unspeakable abomination which seemed to contradict nature, but which in reality merely emphasized her commonest manifestation. The enormous size of those obscene legs was no more than the natural outcome of the sudden overthrow of a delicate balance holding normal growth in check. Destruction of certain glands in the human body, or misguided tampering with their perfectly adjusted excretions, might well result in a corresponding monstrosity of a man. What had suddenly shattered the mechanism of control in this instance? Neither man could guess, although Brown might have suspected, had he known the details of Bork's attempted suicide.

The full sunlight had a horrifying effect on the remains. First the smooth black legs swelled slightly, as it filled with an expanding gas. The crushed legs then stiffened and straightened slightly under the increased pressure, and the skin tightened. Was that horrible fragment trying to live and walk? The pressure increased, and the legs began to glisten as if recovering their vitality. Then the joints of all cracked simultaneously; the eight black husks collapsed under the escaping gas, and all rapidly withered, wrinkling in black decay, like the skin of a perfectly embalmed mummy suddenly exposed to the light and air after centuries in the cold darkness of its hermetic tomb. Within ten seconds only a tangled knot of shriveled wisps of skin remained.

The two men stared into one another's faces.

"Did you see it?" they panted together.

"I don't believe it happened," Crane muttered. "We're both crazy."

"But the smell? It's still pouring out of that door. We must see what is in there."

"Not without a light. I'll lock the door and leave you on guard. If anyone wants to get in tell them we have an experiment going, and say I asked that everybody keep out for twenty-four hours."

With shaking hands Crane locked the door and hurried off to the workshops. In ten minutes he was back, trundling an oxy-acetylene blowpipe with four cylinders of gas, half a dozen globes for the floodlights, and two electric torches. He had recovered sufficiently to minimize the danger.

"I thought we might want to clean up after we've seen what there is to see," he grinned, pointing to the blowpipe. "This will throw a four-foot jet of flame hot enough to scorch the devil himself. How are your nerves? The walk did mine good. After all there must be some simple explanation for whatever has happened. Anything that can be explained is nothing to be afraid of. Well, are you ready? We shall have to stand the smell." They knew that they were in an undiscovered world. Would either admit it? No. To be good sports in one another's eyes, they made light of their anticipated discovery. Such is science, and such is artificial human nature.

While Brown played one flashlight on the lintel of the steel door and spotted the other on the path that Crane must take through the darkness, the latter set his jaw and rolled the truck with the blowpipe up to the gallery of the switches. Then he hastily locked the door and left his key in the lock. The odor had decreased somewhat, owing to the partial airing the laboratory had received while the men stood viewing the dissolution of at least one of the dead enemy, but it was still foul enough. Tying their handkerchiefs over the lower parts of their faces, the two set grimly to work. Their first task was to obliterate every trace of nature's madness; theorizing on its probable cause would come later, if at all. For the moment they realized that nothing mattered but speed in sanitation.

Their first tentative moves were slow and cautious. Although the state of the air seemed to prove that no living thing yet lurked in the darkness, they did not rashly tempt death. Armed with the identical slat of white pine which Bork had used in his blundering attempt to electrocute himself, Crane cleared a path four feet broad while Brown played the flashlights on the hideous things in his companion's way. In all sizes and twisted shapes of death, from balls of black legs no bigger than a rat to contorted monstrosities like enormous jet black spider crabs, the rapidly decaying victims of their own uncontrolled vitality cluttered every yard of steel galleries and cement floors, and depended in loathsome festoons from every railing.

A sudden thud in the darkness to their left, followed by a dry rustle, brought both men to an instant halt, their skins tingling with an unnatural fear. Sweeping his flashlight up to the ceiling, Brown saw what had happened. From every steel girder hundreds of the dead enemy hung in precarious equilibrium. Now and then one swayed slightly, its unstable balance shifting under the rapid progress of a ravenous decay that devoured the softer parts of the enormous body with incredible speed, and tilted the harder remnant of legs and carapace a hairbreadth downward toward the inevitable fall. Underfoot a thick, slippery scum of innumerable black bodies, from the size of wheat grains to mere specks barely visible, marked the sudden slaughter of a self-perpetuating host extinguished in the very act of seizing upon unnatural life.

How had they lived? On what had they fed? The long evil legs were mere distended sacs of skin filled with foul air. Had they swelled to their terrifying dimensions by assimilating the gases of the atmosphere and transmuting the dust particles of the air into the tenuous substance of their skins? It seemed incredible; yet, for the moment, no other explanation even faintly rational suggested itself. Presently Crane turned over two enormous black husks still intertwined in their death embrace. The hollow black fangs of each were sunk deep into the hard thorax of the other. Their first sustenance—whatever may have been its nature—dissipated, the starving brutes had devoured one another.

"I killed a black widow in here yesterday," Crane remarked in a strained voice. "Or was it a thousand years ago? There must be some connection between that one and these. The mark of the red hour glass is on at least half of them. Those without it are the males. Look at those two—the bigger one has the hour glass, the other hasn't. The big one was the female; the other her mate. When she began starving to death, she tried to eat him, only he got his fangs into her, too. Then they both died. Does what I did yesterday explain this nightmare? You're something of a biologist; you ought to have a theory."

"I have," Brown admitted, "but it is worse than this nightmare. Don't go up that ladder. Some of them may still be alive."

Disregarding the doctor's horrified protests, Crane began climbing the vertical steel ladder against the side wall.

"Throw the light up ahead of me," he directed, pitching off two dangling carcasses from the fifth rung. "I've got to get up to the floodlights and stick in some new globes. Two will do."

He made his perilous way up to the steel girders, kicked a footway free for himself along the broadest, and coolly inserted a new light globe. As he screwed home the bulb, the light flashed on, revealing for the first time the full horror of that black shambles like a madman's dream of hell. Brown vented an involuntary shout, and Crane for a moment tottered as if about to lose his balance. Recovering, he walked coolly along the cross girder and screwed in the second bulb, about fifty feet from the second and directly over the pit of the transformers.

"Is that enough light?" he called down.

"Too much. I mean, I don't want to see it."

"You've got to, until we clean up this mess from floor to ceiling. While I'm up here, I might as well clear the rafters. You get a board and begin sweeping them into piles. There's a broom in the janitor's cupboard over there to the left."

For ten horrible hours they toiled in the noisome air of that nightmare tomb, sweeping the twisted black abominations into stacks, and applying the fierce white jet of the oxy-acetylene torch to each the moment it was ready. Nor did they neglect to spray the withering fire over every inch of the concrete floor. Some spark of vitality might still linger in the fine black sand of innumerable eggs, that had burst like capsules of ripe poppy seed from the bodies of the dead females.

At last, shortly after two in the morning, their gruesome task came to an end. Both men were sick from the foul nauseating fumes, and exhausted of mind by their protracted battle with an enemy that had defied nature only to expire hideously.

"I dare not go to bed with that still in my eyes," Brown confessed.

"Nor I," Crane admitted. "Let's take a long walk and blow our lungs clean."

"That suits me. By the way, how is your skin?"

"Itching again. But I can stand it till morning. Anything would be bearable after that nightmare."

They emerged into the cool night air and filled their lungs.

"I never knew that air could taste like this," Brown sighed, exhaling and breathing deeply again. "Aren't you going to lock the door?"

"No. For once I'm going to break Kent's pet rule. This place must smell clean by morning."

"What about your tube? Somebody may come and tamper with it."

"That's so," Crane agreed. "Perhaps I had better lock up after all."

He reentered the laboratory and turned on the lights.

"The tube's gone!" he shouted. So engrossed had he been in the business of destroying the enemy that until this moment he had not noticed his loss. "There's nothing left of it but the concrete stand."

Brown followed him down the steel stairs to investigate. They found nothing that threw any light on the mysterious disappearance of the two-million-volt tube. Only a vitrified white patch on the flat top of the concrete stand hinted at some unusual disaster. An outgush of transcendent heat had fused the concrete into a glassy pillar. Glancing up, Crane saw the melted remnants of the connecting wires dangling from their support.

"Short circuited by some fool's carelessness," he muttered. "Whose?"

"Bork's?" the doctor suggested. "He would be the only man

likely to experiment with your tube. None of the others have worked with it, have they?"

"No. Bork was the only man beside myself who ever touched it. If he did this, it was deliberate. No wonder he committed suicide. Probably he meant just to set me back a month or two and blew out the whole thing. Bork was always a bungler. This is what I get for trying to make a man out of a fool. Never again!"

They retraced their steps to the gallery of the switches.

"Look at that," Crane exclaimed, pointing to the long row of ebony handles pressed securely home. "He short circuited the whole battery of transformers, too. It will take a month to repair that fool's damage. He must have been drunker than usual. Then he committed suicide to escape going to the penitentiary. Well, he did one sensible thing in his life."

"What will you tell Kent now?" the doctor asked as they again emerged into the fresh air, leaving the door wide open. "Will you still stick up for Bork?"

"Why not? Calling him what he was won't restore my tube. And I shouldn't care if it did. Tomorrow I begin work on the twenty million volt tube. Someone will find the door open in the morning and report the damage to Kent. Then the papers will theorize that some enemy of the Foundation stole a key to the laboratory and did ten thousand dollars worth of damage at one swipe. I'll not contradict them."

They walked till sunrise, trying to purge their eyes and minds of the night's horror. The disappearance of the tube gave Brown a further clue to the mystery, but he did not confide it to Crane. Until he could learn more of the action of extremely short waves on living tissue he would keep his daring hypotheses to himself. In the meantime there was one simple check which he could easily apply.

Just as they sat down for coffee at an all-night lunch counter, Brown had a sudden thought which filled him with alarm.

"What did you do with the water in the bathtub after I left?"

"Let it down the drain, of course," Crane replied. "What else was there to do?"

"Nothing, I suppose," Brown admitted. "Only I wish you hadn't. It is probably diffusing into the salt water of the bay by now."

"You got a whole bottleful," Crane reminded him. "Wasn't that enough?"

"Plenty," the doctor muttered.

On reaching his house, the doctor carefully packed the bottle of water in his black bag and hurried with it to call on one of his friends, Professor Wilkes, a specialist at the university in the protozoa. Wilkes was one of those fairly venerable scientists who live on the reputations of their prime, do nothing, and look down their noses at younger, more aggressive investigators who accomplish something. His air was that of a once-nimble sand flea soured by experience; his once flaming hair had gone dull reddish streaked with gray, and his lean, angular body was a habitual protest against the radicalism—scientific—of the younger generation. The professor was just sitting down to breakfast when Brown burst in on him.

"Check me up on this, will you?" he began without preliminaries of any kind. "Either I'm losing my mind or this water is alive with microscopic protozoa new to biology. Until last night I wasn't sure of my guess—I searched all the books, but wasn't convinced by not finding any of these described. I might have overlooked known species mentioned only in out-of-the-way papers. Last night settled it. These things *must* be new—to the extent at least of being radical mutations from known species. Their life cycle is entirely different from anything yet described."

With a curious glance at his friend's face, Professor Wilkes abandoned his breakfast, gravely took the bottle of water, and preceded Brown into the study. Having carefully prepared a slide with a drop of the miraculous water, he then applied his eye to the lens and slowly adjusted the focus. For a full minute there was a tense silence, broken only by Brown's unsuccessful efforts to breathe naturally. At last the professor glanced up.

"Are you sure you have brought me the right sample?"

"Positive. Aren't they new species?"

"Look for yourself," the professor invited, rising and making way for his excited friend.

One look was enough for Brown. With a short exclamation he straightened up and rubbed his eyes.

"Have I dreamt it all?"

"Probably," Wilkes remarked drily. "Alter the focus to suit yourself. Take a good, steady look."

Brown did so, peering into the tiny speck of moisture which concealed his imagined discovery. In silence he prepared half a dozen slides and subjected them to the same pitiless scrutiny.

"I was mistaken," he grudgingly admitted at last. "That water is completely sterile. Unnaturally sterile," he added after an awkward pause, in which he reddened uncomfortably under the professor's sympathetic regard. "There's not a trace of organic matter in it."

"You boiled it and filtered it through porcelain?" Wilkes suggested kindly.

"If so, I don't recall having done so. In fact, I'm certain I did not."

"Perhaps your protozoa all dissolved of themselves," the professor hinted, with just a trace of sarcasm.

"Could *these* dissolve?" Brown demanded, suddenly exhibiting a sheaf of the drawings he had made.

The professor silently took the sketches and stood shuffling them through his long fingers, occasionally pausing with a faint smile to admire the imaginative beauty of some particularly exotic "animal". Without a word he handed them back. His manner plainly intimated that the interview, so far as it concerned him, was at an end.

"You think I never saw the originals of those?" Brown protested.

"My dear doctor, I think nothing whatever about them. My advice to you is to go home, go to bed, and stay there for a week. You have been over-exerting your mind."

Brown restored the despised sketches to his pocket.

"Would you ask one of your colleagues in the department of chemistry to analyze the water in this bottle if I leave it with you?"

The professor agreed good-naturedly, and Brown left him to finish his belated breakfast. The moment his eccentric visitor was safely down the steps, Wilkes carried the mysterious bottle into the kitchen, thoughtfully extracted the cork, and poured the contents down the sink. Then he threw the bottle into the waste can.

"Mad as a hatter," he remarked. "Poor Brown! It's a blessing he has no wife."

The professor's theory was partly confirmed by an amusing item in the morning paper. This was a hilarious account of the doctor's progress through the city with the voluble Bertha as his only companion. The writer concluded his graphic description with the hint that Doctor Brown, the hard-shelled bachelor, intended his triumphal ride as a gentle hint to the ladies of Seattle that he preferred the company of a fussy hen to their own.

"Mad," the professor repeated to himself. "I'm glad I never consulted him."

Crane was just about to retire after a thorough disinfecting, when the telephone rang. It was Brown.

"We actually did all that last night?" the doctor's perturbed voice inquired.

"I don't know what you mean by 'that', but I can guess. We did, if you have in mind what I have."

"Is it anything about spiders?"

"You might call them that."

"So it was real?"

"Real? I'll say it was. Until you loosen up and explain how it could ever happen, I'm going to think of it as the black widow's revenge. You said it fitted some theory of your own. Come over this afternoon and save my mind. I'll need you."

"Who is going to take care of me?" Brown demanded. "All that we did last night is nothing to what I've just done. I've proved myself a hopeless lunatic to the worst old gossip on the university faculty. It will be all over town by tonight.

Crane chuckled.

"We did such a thorough job in cleaning up that nobody will ever believe a word we say, if we let the least hint escape. You won't catch me letting it out. Better follow my example and sleep it off."

"I will, as soon as I have given Bertha her breakfast. By the way, what is the proper thing to feed a hen in the morning?"

"Spiders."

5

HIS JOKE

FROM THE MOMENT THE LIBRARIAN LEFT De Soto alone with the electrical books till eleven o'clock at night, when the closing bell rang, the new man concentrated every ounce of his tremendous vitality on his self-appointed task. Had he been told that human beings—except a few of the most highly gifted—master the printed page a line or a paragraph at a time, he would have laughed incredulously. He himself had "always" digested the information in books by turning their pages as fast as his nimble fingers would let him, and taking in each page at a glance.

The first books, purely descriptive, that he photographed in this manner on his mind, irritated him almost beyond endurance. Why did the writers go to such tedious length to state what was trivially obvious? De Soto began to conceive a mild contempt for the science of electricity as expounded in college texts and popular treatises. In some indefinable way it all seemed an old legend dimly remembered from a forgotten life. The rudimentary knowledge of the universal forces of nature was as instinctive in him as breathing. Had he not 'always' recognized instantly the hidden interplay of natural things as intuitively as he perceived the noonday sun? Why then did these tiresome authors throw up an endless dust of words and irrelevant theories between themselves and the truth of nature that anyone but a blind idiot could apprehend? The amazing speed of his own vital processes made the laboriously acquired knowledge of generations of scientific men seem deadly slow and wilfully blundersome. Why could they not open their eyes and see what lay all about them?

It was only with the thirtieth book hastily sampled that De Soto's naive conceit received a salutary check. He had just flashed through the massive bulk of Faraday's monumental "Experimental Researches", marveling at the man's painstaking labor to expound the obvious, when he encountered a new language, written in bizarre symbols, of which he could make out nothing. Exasperated by his failure to understand the writer's hieroglyphics, he glanced at the back of the cover to learn the author's name. It was Clerk Maxwell, and the title of the treatise was "Electricity and Magnetism, Vol. 1." De Soto rose from the table and went in search of the reference librarian. She greeted him at her desk with a welcoming smile.

"Are you finding what you want?"

"Some of it. Can you tell me what language this is?"

She glanced at the beautifully printed page and laughed.

"Higher mathematics, I should judge from its appearance."

"Have you any books on this sort of thing?"

"Several hundred. I'll show you where they are."

She left him to his own devices in a stack of shelf upon shelf of assorted mathematics, from beginners' arithmetics to appalling tomes on mathematical physics that were consulted, on the average, perhaps, once in two years by the patrons of the library.

"Do I have to read all these?" he muttered, turning the diagrammed pages of a descriptive geometry. "More stuff that need never have been printed. Why do they write it? Couldn't an idiot see that this is all so?"

In spite of himself, as he worked his lightning way steadily through modern higher algebra, analytic geometry, the calculus, and the theory of functions of real and complex variables, he began to become interested. Here at last was the simple, adequate language of nature herself. It was terse and luminously expressive in a highly suggestive way—unlike the ton or so of solid prose he had already digested against his will. What the italicized theorems left unsaid frequently expressed more than they purported to tell. De Soto found his own intelligence leaping ahead of the printed formulas, and revelling in the automatic interplay of the concepts their brevity suggested.

Gradually a strange, new light dawned on him. This beautiful language after all was but another shovelful of unnecessary dust thrown up by clumsy workers between themselves and nature. Why go to all this fuss to torture and disguise the obvious? Why not look ahead, and in one swift glance see the beginning and the end of every laborious, unnecessary demonstration, as but different aspects of one self-evident truth? All these imposing regiments of equations and diagrams, that marched and countermarched endlessly through book after book, were merely the fickle mercenaries of men too indolent to win their own battles. By a conscious exercise of its innate power the mind, if only it let itself go, might perceive nature itself and not this pale allegory of halting symbols. Did the writers of scientific books need all these lumbering aids to direct comprehension?

"The world must be full of idiots," De Soto sighed simply, putting back a profound treatise on the partial differential equations of physics. "Has it always been so? I can't seem to remember a time when I didn't know all this stuff by instinct. Still, as I have to live in the world, I must learn to speak its silly language."

There was nothing miraculous about De Soto's performance. A profound physical change in the structure of every cell in his body had accelerated his rate of living—or at least of thinking and perceiving—many thousandfold beyond that of any human being that has yet been evolved. He had not waited for evolution. A million years hence the whole race will no doubt have passed the point which he, by a blundering accident, attained in the billionth of a second. Whether language, mathematical or other, will survive to plague our descendants of the year 1,000,000, is doubtful. These feeble aids will have become as useless as the meaningless magic of our remote ancestors. De Soto was but a partial, accidental anticipation of the more sophisticated and yet more natural race into which time and the secular flux of chance are slowly transforming our kind. Viewing the vast accumulations of lore which he had absorbed and spontaneously outgrown, De Soto felt old and depressed. What could he do in a world that still tripped itself at every step on its swaddling clothes? Although he did not realize what he was, he felt a chilling sense of poverty and isolation. Sobered in his exultant vitality, he turned slowly back to resume his pursuit of the mysteries of matter. He began where he had left off, with Maxwell's treatise. This was now as childish as the mathematics he had outgrown. Book after book of high speculation and curiously distorted fact passed under his lightning scrutiny, was mastered for what its author intended it to be, and tossed aside.

"Wrong, at bottom, every last one of them," was his somewhat presumptuous verdict as he closed the last, a modern masterpiece on theories of quanta and radiation. "Why will they not see what stares them in the face? The universe lies all about them, everywhere, and like impossible contortionists in an insane circus they succeeded in turning their back on all quarters of it at once."

It was no harsher a verdict than many a man of science would pass today on the science of the Greeks, or even on that of three centuries ago. Are we as blind as De Soto imagined we are?

"What can I do?" he asked aloud. "If these are the problems they try to solve, how will they ever understand a real one?"

The jangle of the closing bell cut short his gloomy meditations, and he walked slowly out of the building. The crisp night air brought the blood tingling to his cheeks and forehead in a surge of stimulated vitality. Immediately he felt young again, and walked briskly down the brightly lighted boulevard leading to the civic center. The tide of night traffic brought some all but extinct memory of a former existence into momentary life. For one awful second he doubted his identity.

"Who am I?" he gasped, stopping abruptly before the plate glass window of a soda palace. "Am I insane?"

He caught his own reflection in one of the decorative mirrors of the window. It stared back at him, ruddy-lipped, swarthy-skinned, black-eyed and, above all, young with an air of perpetual youth.

"That is not my face," he muttered. "I was never like that. His hair is jet black. Mine is—." He stopped, unable to continue. "That man," he whispered, "is looking back on me from the farthest side of a grave where I was buried a million years ago—or where I am to be buried a million years hence. It is all the same. I am dead and buried, and yet I live."

The 'lost' feeling dispersed as quickly and as mysteriously as it had come. De Soto turned from the window and walked with springy step toward a small park. Although it was now nearly twenty-four hours since he had tasted food, he experienced no hunger. He did, however, feel the need of sleep. Where should he lie down and rest? A glance at the unfathomable vault of the sapphire sky, ablaze with steady stars, convinced him that even the airiest room on such a night would be intolerably stuffy. His problem solved itself. A vacant bench under a fragrant chestnut tree, whose leaves rustled mysteriously in the soft breeze, invited him to rest. In five minutes he was fast asleep. The policeman in the park padded by as noiselessly as a cat, cast the sleeping young man a cynical glance, decided he was broke but sober, and passed on, leaving him in peace.

De Soto slept about four hours, an even, dreamless sleep of complete refreshment. Waking fully and instantly shortly after four o'clock, he felt alive with energy and ready for a long day's work. It was still dark, without a hint of the coming dawn. The library would not open for nearly five hours yet. Suddenly De Soto realized that the library was not his goal, and never would be again. Of what value were all its dusty mountains of dead knowledge? He had mastered the best of its scientific offerings. If the rest—literature, philosophy, and art—was of no greater merit relatively than the cream of the science, it would not interest him. It was not worth inspecting. All of it must be like the science—the first, awkward effort of a race, that had discovered its mind but yesterday, trying to grasp the meaning of life, and failing ludicrously in the attempt. Libraries and all they signified belonged definitely to his irrevocable past—the gray age of a million years ago.

Hunger asserting itself, he rose to seek a clean eating place. He found what he wanted on the cross street opposite the park, a small place, but spotlessly clean.

Only the cook and one waiter were on duty. Neither gave him more than a casual inspection as he entered, for both were servants to the core without one spark of imagination to lighten their completely bovine lives. To such human beings all others appear as listless as themselves. The food was well cooked and neatly served. On these scores there was no complaint. Nevertheless, as De Soto sat sipping his final glass of water, he experienced a vague feeling of discontent. What had aroused his indolent animosity? Chancing to meet the waiter's eyes, he knew. It was the waiter and the cook.

"A pair of mistakes," he thought to himself. "What does either get out of life? They might as well be vegetables. Neither has any interest in his work or in his life. Why don't they do something different? Or why," he thought grimly after a moment's reflection, "don't they hang themselves? The cooks, the waiters, the manager and all at that place yesterday morning were different. They were alive, and enjoyed life—in their own way. But still they were enjoying it. That is the great point. These two are dead and lack the genius to wish they were buried."

Without tipping the moribund waiter, De Soto paid his bill and left the place in disgust. His harsh judgment on the sad-eyed waiter and the harmless, bored cook was of a piece with his estimate of modern physics. They and it were alike useless, both to themselves and to any rational society.

Not only mentally but also physically De Soto was an entirely different being from the stupid, unhealthy Bork in whom he had originated. It is therefore no exaggeration to say that De Soto was only about twenty-four hours old when he left the discouraged cook and waiter, and stepped out into the cool, bracing air of the early morning. In appearance he was a strikingly handsome youth of twenty, with almost a preternatural intelligence shining from his black eyes and glowing from every feature. There was, however, a haunting 'something' about his whole expression which contradicted his vivid youthfulness. An elusive seriousness belied the faintly smiling lips, and a still less tangible shadow of extreme old age lurked behind the light shining from his eyes. It was as if he had seen everything that life on this planet will have to offer for the next ten thousand centuries and, having seen it, was ironically disillusioned by its meaningless futility. Another man having had a similar vision might have been lifted to ecstasy over the lightning progress of the human race; not so De Soto. It was merely a matter of temperament; the dead Bork had not been completely burned out of the living man.

"What shall I do?" he pondered, as his rapid stride hurried him through the darkness. "Is anything worth doing in a world like this? If everything seems stale to me, how can I make it appear fresh and desirable to others? There are too many of them—millions and millions and millions like that cook and waiter. They take everything and give nothing. Give them nothing and they ask for nothing, provided they be permitted to exist. Why permit them?" he continued coldly. Then, after a long blank in which he neither thought nor felt, his lips silently framed the unanswerable question, "Why should any human beings live?" The obvious retort flashed into his mind, "because they can and because they do." That, however, was not an answer to the 'why' as he meant it. His purpose was taking shape. Before many hours he was to decide what he should do with his inexhaustible health and his boundless talent. Rather, the transformed cells of his body were to decide. It was a decision such as our own distant descendants may reach some day—the verdict of an incredibly old and sophisticated man infinitely disillusioned.

He stopped abruptly in his walk, hypnotized by the strange familiarity of a massive, rectangular building, which loomed up forbiddingly before him in the graying darkness. Where had he seen such a building before? As if threading the shadowy mazes of a previous existence in a dream, he stole toward a sharp oblong of sheer black on the dimly visible wall. The door was open. Before he realized what he was doing, he had entered and turned the switch of the floodlights.

"Where have I seen this place before?" he muttered staring down into the pit of transformers. Another memory struggled up from the wreckage of lost associations, but he could not place it. "There should be twelve lights up there, not only two."

His feet urged him to descend the steel stairway to the pit. The gigantic gray devils towering up on their rigid legs were familiar enough in a subtle way; he had seen pictures of similar monsters in the books at the library. Intuitively and from his comprehensive reading he knew immediately their evil powers and their uses.

"These are unnecessarily big and complicated," he remarked

aloud, as if giving his considered estimate to some attentive listener. "Don't you see that you could build a single, compact one to do all that these forty can? All you need—" He launched into a rapid description, bristling with technicalities, of what was necessary for his projected improvements. For an hour and a half he roamed through the laboratory, examining every piece of apparatus, criticizing and contemptuously condemning each in turn. "A hopeless bungle," was his final comment as he ascended the stairs to the gallery of the switches.

Daylight was now streaming coldly in through the open door. De Soto walked to the switch to turn off the lights. The main switchboard controlling the transformers caught his eye, and he noted that the long row of ebonite handles were all down as far as they would go. This was no condition to leave a switchboard in, no matter if it was "dead."

As he opened the switches, his vision included the long bench beneath them. An empty cigar box lay open on the bench. De Soto picked it up, turned it over and over in his hands and finally set it down, his mind vaguely unsatisfied. It recalled nothing to him. Yet he had a chilly feeling that some incident connected with that box had marked the turning point of his life.

With a sigh he left the laboratory.

"Where was I born?" he pondered, gazing up at the golden flush which presaged the rising sun. "When? Was it in Buenos Aires, as I told that woman in the library? How strange that I should have forgotten everything of my early life! This is what they call amnesia, I suppose. Well, what does it matter so long as I know who I am now? Yes, I must have forgotten. Let it go."

A neat inscription over the entrance to the building opposite him announced that this was the Erickson Foundation for Electrical Research, and a small tablet on the upper step added Administration Building.

"This is where I intended to ask for work," he remembered, as from a past inconceivably remote. Between this dawning day and its yesterday, when he had scanned almost in a glance the sum total of existing physical knowledge, lay an eternity in his maturing mind. The twenty-four hours had aged him so that he looked back on his yesterday's ambition as the uninformed dream of an eager child. Although he knew better now than to pursue his childish purpose—whatever it may have been—he adhered to his plan of working at the Foundation. Already his initial intention was more than half forgotten; his new purpose, he thought, would at least help him to pass the time.

No sooner was this resolve formulated than a queer echo made itself heard from the deepest recesses of his mind. "Will you pass the time," the doubter whispered, "or will the time pass you?" For a moment a chill conviction seized him that both possibilities suggested by his subconscious mind were to be fulfilled to the last letter, and that he was merely a helpless drifter on a black ocean without shore or tide. He quickened his pace to ward off the fingering chill that pattered over his whole body, determined to lose himself in vigorous exercise until he could call upon the director and ask for work.

While De Soto was walking off his depression, Brown was endeavoring to sleep. All he achieved was a fitful nightmare till about ten o'clock in the morning. Giving up the attempt, he rose and made a deliberate effort to resume his normal habits. This morning should be like any other of his orderly life. He bathed, shaved and rang for the housekeeper to order breakfast. The doctor was an early riser—when he had not been up all night—and his first two hours out of bed were his time for loafing. No man, he was wont to assert, could hope to wake up fully in less than two hours. His period of relaxation was not wholly wasted, however, for he usually managed to read at least one article in the current scientific journals before going to work. For the moment he was determined to forget his nightmare in the laboratory and his own vanishing protozoa.

The morning paper was rather duller than usual, and Brown was just about to discard it in favor of the "Biological Review," when an unobtrusive paragraph on the last page caught his eye. "Fishermen report strange malady," the caption ran. The fish, Brown learned on reading the article, not the fishermen, were the sufferers. Moreover, only salt water fish were affected. The disease manifested itself in discolored blotches of all hues—blue, green, yellow, purple and red—on the skin and fins. The flesh seemed as firm and sound as ever. Probably, the report stated, the discolorations were harmless.

"Perhaps," Brown agreed, cutting out the paragraph. "More probably they are not. This must be looked into by the Board of Health. If the fish trust thinks it can get away with anything like this, it is badly mistaken. That story should have been printed on the front page."

A short conversation over the telephone set the appropriate machinery into instant motion. The Chief of Health promised to send out his squads at once to seize and destroy all spotted fish exposed for sale.

"That settles that," Brown remarked, picking up the "Biological Review." But was it settled? Only salt water fish were being affected. The doctor meditatively extracted the sketches of his despised protozoa from an inner pocket and stood thoughtfully regarding them. "Are they as mad as the professor thought? What if they are? Discoveries aren't made by 'safety first.' I'm going to do that experiment the minute Crane gets his twenty million tube built and going." He rang for the housekeeper.

"Will you make it part of your work," he requested the capable woman who responded, "to see that Bertha is well taken care of? It may be weeks until I have any use for her, but in the meantime I want her to hatch all the eggs she lays."

The housekeeper put a broad hand over her mouth and turned aside. When she recovered her composure, she ventured a practical suggestion.

"You might get Bertha a husband, if you want her eggs to hatch."

"Of course," the doctor agreed hastily. "Will you see to it?"

When she had retired out of earshot, Brown called up Crane's apartment.

"Hullo, Crane? Could you sleep? No more could I. How is your skin this morning?"

"Practically better. I can't understand——"

"Neither can I, yet. But I'm getting warm. Have a look at the last page of this morning's *Sun.* There's an interesting article on fish. By the way, when you take a bath hereafter, sterilize the water as well as you can before letting it down the drain. Better get a liberal supply of cyanide of mercury and put about a tumblerful into the tub when you are through."

"What for?"

"General precaution. That amount won't do any harm by the time it reaches the sea, and it may prevent a world of mischief before it gets there."

"Am I dangerously infected?"

"Probably not, in any mundane way. That's what makes you so interesting. Going to start work on your new tube this morning?"

"Yes, as soon as I have broken the news to Kent that he must spend about ten thousand dollars putting the transformers right again. Friend Bork did a rare job as his parting shot. I was just going down to the laboratory when you called up. By the way, never tell a soul about last night. I'm beginning to believe it never happened."

"I'm not advertising. If anything interesting happens, let me know."

On reaching Kent's outer office, Crane was greeted with ominous formality by the secretary.

"Will you take a seat, Doctor Crane," she said, "until Mr. Kent is ready to see you? He was about to send for you."

"Who's dead now?" Crane inquired flippantly.

The secretary ignored Crane's levity and merely stated that the director was "in conference" with the trustees.

"I guess I'm fired," Crane remarked to the ceiling. "If so, will you please tell Mr. Kent that I resigned last night?"

As the secretary deigned no reply, Crane moodily sat down and lost himself in brooding. The spectacular damage to the transformers, he guessed, was the inspiration of Kent's untimely session with the trustees. They would naturally blame him for having left the door unlocked. Well, he could not prove that he hadn't. Let them do what they liked; he didn't give a damn. In spite of his assumed indifference Crane realized that a dishonorable discharge from the Foundation would cut pretty deeply into his self-respect. Moreover, now that he was about to be fired—as he imagined—he suddenly conceived a warm affection for the Foundation and for every member of its staff, except of course Kent. A resonant voice asking for an appointment with the director caused him to look up.

The secretary was staring in fascination at the dark, intelligent face of the man addressing her, unable, apparently, to follow his question. Like the librarian she was wondering where she had seen this striking man before, although she was in her early twenties with no disastrous love affairs in her past.

"Will you please make an appointment for me with the director?" that elusively musical voice repeated.

"As soon as Mr. Kent is free," she murmured, "I know he will be delighted to see you, Mr.——"

"De Soto."

"Will you wait here? Mr. Kent is in conference."

"Must I wait? I should prefer to see him at once, as I have a full day ahead."

"I'll see," she volunteered. Then it occurred to her that she did not know the young man's mission.

"Mr. Kent will ask me your business." She all but apologized for the indelicacy.

De Soto unfolded a morning extra and indicated the joyous headline: "ERICKSON LABORATORY DESTROYED." Of course the laboratory was not destroyed; the giant transformers only were disabled. To have stated so in cold print would have killed the story. On this occasion neither the director nor the trustees were reluctant to confide their misfortune to the press. The janitor early discovered the open door; a short investigation by members of the staff, summoned from the workshops, disclosed the extent of the damage. Too obviously it was the work of an enemy from the inside. Kent, at the moment of De Soto's appearance, was endeavoring to convince the trustees that Crane was guilty.

"He practically told me to go to hell," he vociferated. "And he said he cared nothing for what he called his 'silly job.' "

"Why not ask your secretary to call him, Mr. Kent, and ask him whether he did it?"

Kent was game. He pressed a button. The buzzer called just as De Soto unfolded the extra for the secretary's gaze.

"I have called about the damage to the transformers," he explained. "Please tell Mr. Kent that."

"I will," she promised, and hastened to answer the insistent buzzer.

All the time that De Soto stood parleying with the secretary, Crane studied him minutely. By one of those common but unaccountable quirks of human nature that often mystify us, he took an immediate and violent dislike to the swarthy young man with the peculiar voice. Possibly Crane's own frigid reception by the secretary, contrasted with De Soto's, may have touched the hair trigger of his masculine jealousy and self-love, although the secretary was nothing to him. Ordinarily he treated her as a pretty piece of furniture. Whatever the cause of Crane's instinctive dislike, the feeling itself was not to be denied. That it was primitive and irrational but made it more significant. Crane followed the young man's every movement and listened avidly to every changing inflection of his voice. Had he been capable of self-analysis he would have summed up his conclusions somewhat thus: "That young man impresses me as a thoroughly bad egg. I'm going to watch him."

When De Soto asserted that he had called about the damage to the transformers, Crane's curiosity naturally was aroused to the point of acute physical discomfort. Left alone for a few minutes with the young man, Crane decided to break the ice.

"You are an electrician?" he demanded of De Soto's back.

De Soto wheeled about sharply and found himself looking up at a rather grim, square-jawed face on which more than a hint of hostility showed. Before he knew what he was doing, De Soto found himself rocking in uncontrollable laughter. Do what he would to stop his rude mirth, he failed. What was the joke? Why did the man's face awaken the haunting sense of the ironically ridiculous? No ghost of a memory whispered to remind De Soto of his past. Yet, what he planned to do with his unbounded talents for the good of humanity seemed irresistibly ludicrous, and the humor of the situation was focussed in some mysterious way on the disconcerted face looking down into his own. Crane stood it as long as he could.

"If you'll tell me what's funny, I'll laugh too. Go on; don't mind me."

De Soto gulped and subsided—outwardly.

"I was laughing at you taking me for an electrician," he explained glibly. "In a way I am, although radiations of the shortest wavelengths—ultra violet, X-rays, gamma rays, and so on, clear into the region of the hardest cosmic rays—are my specialties. You see, I have graduated from electrical engineering. That is what you thought I was interested in, from my remark to the secretary about the transformers. That's merely to help me into a job here."

"You say you have worked in the cosmic rays, and even beyond?" Crane demanded suspiciously, scenting a quack in this plausible young man. "I don't seem to recall the name of De Soto in that field. Where did you do your work, if I may ask?"

De Soto submitted to the unwarranted cross-examination with a good grace.

"In Buenos Aires."

"At the Universidad Nacional de Buenos Aires?" Crane suggested skeptically. He knew perfectly well that the national university of the Argentine offered no work in the field De Soto claimed as his own. De Soto, however, whether from accident or design, side-stepped the rather obvious trap. Unknown to himself, he was lying, and lying quite ably. The hastily imagined fiction of his birth in Buenos Aires, which he had invented but a few short hours previously, was already a fixed part of the life he had "always" known.

"No," he replied. "I did all my studying privately."

"Self taught?" Crane suggested in much the same bantering tone he might have used to Bork. De Soto ignored the hidden slur.

"Why not? What are books for?"

"So you never have worked in a laboratory?"

For a moment De Soto was at a loss. He *had* carried out electrical experiments. But where? Somehow the sure feeling of familiarity with electrical apparatus, of which he felt confident, did not date with the rest of his self-acquired education. Doubtless his inability to connect the two was but another trifling instance of the amnesia from which he seemed to suffer. Trusting that he was telling the truth, he gave what appeared to him at the moment as the only reasonable answer.

"I had a small laboratory of my own."

Crane received this in silence. Further cross-examination was postponed by the return of the secretary. She had been urging De Soto's priority over Crane's in the matter of an interview with Kent and the trustees.

"Mr. Kent will see you now," she announced, nodding to De Soto.

Followed by the suspicious eyes of Crane, De Soto disappeared into the inner sanctum. Crane favored the secretary with a sour look and resumed his chair. Whether they fired him or not, he was not going to leave until he learned De Soto's object.

Once in the director's sanctum, De Soto found himself the mark for seven pairs of questioning eyes. Kent rose from his seat at the end of the long table, his hand extended in formal welcome, while the six trustees turned in their chairs to get a better look at the newcomer.

"You are from the press, Mr. De Soto?" Kent inquired.

"No," De Soto laughed. "I came about this." He tapped the article about the damaged transformers. "The papers say there is a hundred thousand dollars worth of mischief done."

"Exaggerated," one of the trustees interrupted. "Ten thousand will cover it."

"What I propose is this," De Soto continued, acknowledging the trustee's remark with a slight nod. "For not more than a total outlay of five thousand dollars you can duplicate your whole battery of transformers. Now, this is what I suggest. Spend five thousand dollars, and I will show you how to do everything your high tension laboratory ever did, or ever could do—provided you were to repair the damage. Moreover, the entire apparatus will not exceed a cubic yard in bulk."

Kent glanced at the trustees with a wan smile. His irreproachable secretary, departing from her habitual caution, had admitted an impossible crank to disrupt a most important conference. The quality of Kent's smile slowly changed. It became puzzled. Why, precisely, had he and the trustees listened so attentively to this swarthy young man's preposterous claims? A trustee cut the knot.

"Can you make good on what you propose?" In spite of his shrewd business sense he was strongly attracted by this magnetic young fellow's personality and the strangely luminous, rational glow of his eyes. "If you are not talking nonsense, you should be able to convince experts. Want the chance?"

"Experts such as you have," De Soto replied truthfully but with unintentional rudeness, "probably wouldn't understand my plans. However, if you wish, I will try to explain."

"Ring for Doctor Crane,' 'the trustee snapped. "I believe Mr. De Soto knows what he is talking about."

At the mention of Crane's name, De Soto controlled a strong impulse to laugh. When Crane entered, glum as a naughty boy expecting a reprimand, De Soto turned his back on him. All through the ensuing discussion the man who had been Bork never once looked his former chief in the face. Some uneradicated instinct held him back, although consciously he had not the shadow of a memory of his former enemy and would-be friend.

The lively talk began with a trustee's suspicious questioning of De Soto regarding the young man's scientific training. De Soto repeated smoothly what he had already palmed off on Crane, adding several circumstantial details. The story, strange as it seemed, hung together. The trustee, unconvinced, let it pass at its face value. After all, what did it matter who or what the man was, provided he concealed an inexhaustible mine of diamonds under his thick black hair? If he could do what he claimed in the matter of the transformers, this attractive young man must be an inventive genius of the very first rank. At the rate of one dollar apiece for all patents taken out while a member of the staff, De Soto might well be worth several hundred million dollars to the Erickson Foundation before his contract expired. The worldly wise trustee was already rehearsing the terms of the ten-year contract which would sew up De Soto tighter than any dead sailor about to be slipped overboard ever was.

From personalities the discussion soon launched into a bewildering debate of technicalities between De Soto and Crane on the design of high tension apparatus, the construction of more efficient insulators, thinner than the thinnest tissue paper and, most important of all, an entirely novel process for attaining a perfect vacuum. Even to the untechnical trustees and the unscientific director it was evident that the best of the argument was De Soto's from start to finish. As Crane interposed one objection after another, only to have each demolished in turn by a short sentence backed up by a shorter mental calculation, De Soto began to lose patience with the slowness of the expert mind. Finally, turning to the director in exasperation, he asked whether there was one—only one—competent physicist in the Foundation. Kent silently pressed a button. To the worshipping secretary he explained that Mr. De Soto wished to confer with Doctors So-and-So, naming the cream of the specialists enslaved to the Foundation.

They dribbled in by twos and threes, until a full dozen found themselves involved in the most terrific battle of their distinguished scientific careers. By ones and twos they were eliminated temporarily by hard facts or harder formulas hurled at their heads, only to rally for the next attack and be again knocked flat. Whether De Soto was right on his new theories, or on his novel project as a whole, was beyond their powers to decide. There was no doubt, however, that he had all the classical theories and current experiments—old stuff, in his contemptuous phrase—at his finger tips. Lunch time had passed long since; the dinner hour came and went unnoticed, and still the battle raged. At last, shortly before midnight, De Soto had sunk his last opponent. On many of the details of his project they were still unconvinced, but they had shot their last round.

The signing of the contract followed as a matter of course. Departing from their invariable custom, the trustees guaranteed to De Soto royalties of one per cent on all inventions patented by him while a member of the staff, in addition to the usual legal dollar. Argument with such a man might prove more costly in the end than graceful submission. They gave in before he could have a chance of offering his talents to some less greedy competitor.

While De Soto was affixing his signature—Miguel De Soto—to the contract, Kent buttonholed the President of the Board, and poisoned that potentate's mind against the defenseless Crane. All of Crane's flippancy in the face of duty, his flagrant disrespect for decent authority, and finally his heinous offense in leaving the door of the high voltage laboratory open, and so causing thousands of dollars worth of damage, were poured into the President's exhausted ear.

"You want me to ask for his resignation?" the President suggested. "Is that what you're driving at? Give the word, and I'll do it. We have just taken on Mr. De Soto at ten thousand a year *plus* those robbing royalties. Why not save Crane's salary? He's of no further use to us that I can see. Mr. De Soto's field includes Crane's, I gather?" Kent nodded. "Then say the word, and I'll give Doctor Crane his walking papers."

"It might cause criticism if it got out," the cautious Kent demurred. He saw his chance of sticking his longest knife into Crane, and he determined to seize it. In fact, he had worked on the President with just this purpose in view. "Why not cause him to resign voluntarily?"

The President all but grinned. "How?"

"Tell him henceforth he is to act as Mr. De Soto's assistant. Say that we all know how proud he will be to work for so great a scientist as this extraordinary young man has proved himself to be. Don't mention a word about salary. Put it on the purely scientific plane of service."

"Leave it to me."

De Soto stood talking to three of the trustees as the President descended upon Crane. Speaking in a loud voice so that none of the technical staff present should miss the obvious moral, the President delivered his honeyed ultimatum. The faces of the twelve men of the staff went blank; those of the trustees beamed. Here was Crane's chance to show his mettle. The technical experts prepared to offer their own resignations the instant Crane refused the insulting offer; the trustees wondered whose fertile mind had conceived this neat method of firing a faithful employee whose services were no longer as profitable as they had been. Crane's immediate response took all parties completely aback.

"Thank you, Mr. President," he replied gravely. "I shall be honored to serve as Mr. De Soto's technical assistant. And I shall never forget your generosity in giving me this opportunity of showing my loyalty to the Foundation and to Science."

Kent's jaw dropped. Crane's rounded speech of acceptance was a first class imitation of what his own might have been under similar circumstances. What did that long, lean devil of a Texan mean by it? Was he forcing them to fire him outright, like men? Well, they weren't such fools. They would put no shovel in his hand to fling mud at the Foundation. So much for Kent. The President shot him a spiteful glance as if to say, "You've made a fool of me. Two can play that trick, as you will find out before long." The members of the staff looked anywhere but at Crane. They would have backed him up to a man. Who would have suspected him of having a yellow streak? So much for Crane's former friends. De Soto extended his hand in friendly congratulations. Crane shook it vigorously. So much for the conventionalities.

What of Crane? He meant every word he said. That he had phrased his acceptance in fatter rhetoric than he usually fancied was merely the luxury he allowed himself of insulting Kent before the trustees. Crane knew exactly what Kent was trying to do, and he dared his petty tyrant to do it like a man. As for the coldness of his scientific friends, he felt that he could endure it as long as they. It was of no great moment. The one thing that sustained him was his instinct that De Soto was evil to the core, with a black, new evil, venomous beyond human experience. Did not any of the others feel what he sensed with every nerve of his body? No, he admitted; they probably did not. The trustees were blinded by the profitable bargain they had just driven with the new man, the technical experts by his scientific brilliance. All, Crane felt, might more sensibly have sold themselves for thirty cents to the devil. Their gain was a fool's. Although he had no definite feeling as to the precise way in which De Soto was a thing of evil, Crane knew that his silent estimate was just. Until he or the Foundation went under, he would stick by the ship and save all he might. His motive was not love for a corporation which had treated him scurvily, but intense dislike, amounting almost to hate, for the dark young man with the piercing eyes who henceforth was to be his driver.

On passing into the outer office, they found Alice waiting to drive her poor, tired father home. Kent seized the opportunity of further humiliating Crane by presenting De Soto to his daughter with a great show of arch-fatherly effusiveness. Crane observed the comedy, nodded curtly to Alice, and strode out of the building. To overtake his departing enemy with a last barbed dart, Kent raised his voice and insisted that De Soto be his guest for a few days, until he found comfortable quarters near the Foundation.

Two minutes later Kent's car overtook Crane. Alice was driving; Kent was spread out on the back seat. De Soto, sitting up in front by Alice, seemed to be progressing rapidly with the director's daughter.

"That fool!" Crane muttered. "Can't he see what De Soto is? I don't mind being snubbed. But it is a bit thick when he uses Alice to do his dirty work. Well, if she is that kind, I'm not a sentimental sap. She can go to——."

An hour later, De Soto stood as still as a rock on the cork mat of the bathroom in Kent's guest suite, staring with unseeing eyes into the mirror. He had just enjoyed a luxurious bath. The director had kindly lent him a suit of pyjamas, smelling faintly of lavender, and a soft shirt and socks of his own for morning. De Soto had explained that his things were in his room at a distant hotel. In the morning he would lay in a complete outfit. For the moment he was lost in thought, or rather in an ecstasy of pure existence which was neither thought nor sensation. The warm bath had stimulated his circulation, and he was, for the time being, a perfect animal and nothing more. No spark of human intelligence kindled the black eyes staring into his own from the plateglass mirror.

Presently he sighed. Realization of his circumstances and of what he had accomplished during the past twenty hours overwhelmed him like a whirlwind of fire. The eyes in the mirror leapt into life, flaming up from the dead blackness of incredible age to the piercing gaze of intelligence incarnate in perpetual youth. His purpose came back to him. No longer merely a faultless animal, he had remembered his humanity and all that he intended for his fellow human beings. He remembered also the trustees and the director, and the bargain they had driven with him. For an instant his face clouded with fierce scorn. Then he began to shake with silent mirth. His secret joke, multiplied a billion-fold, had returned to comfort him.

"Millions and millions and millions of them," he thought, "like that cook and waiter, and millions more like Crane, those trustees, that director, and his daughter."

Still shaking with suppressed laughter he stole into the bedroom as noiselessly as a tiger and went to bed.

6

DISCHARGED

SIX MONTHS AFTER HE BEGAN WORK AT THE Foundation, De Soto found himself world-famous. Although he never read a newspaper, he could scarcely avoid seeing his name at least once a week in the headlines as he passed the newsstands. He had out-edisoned Edison and out-invented all the electrical inventors of the past seventy-five years—according to the press. Remembering that the papers made a hundred thousand dollars worth of damage out of a paltry ten thousand in the matter of the transformers, we may safely discount these early reports to about ten per cent of their face value. When, some months later De Soto began doing things of greater significance for humanity —things that could not be evaluated in terms of dollars and cents —the press was dumb, and for a sufficient reason. As long as the new so-called luxuries and conveniences of living which De Soto created, seemingly in his sleep, inspired the journalistic tongue, reporters and editors were on familiar ground. But when De Soto, tired of playing Aladdin's lamp to millions who rubbed him the wrong way, turned to the higher and more difficult parts of invention, the world simply did not realize what was happening to it.

De Soto's masterpiece was new in human history. To find its peer we should probably have to go back at least as far as the beginning of geologic time. The human race, in De Soto's vaster enterprises, was merely a rather minor indiscretion on the part of mother nature. But for the first six months of his dazzling career as the king of inventors, Miguel De Soto lived up as best he could to what, he sensed, a somewhat pampered world expects from its geniuses. They asked him for bread and he gave them cake. The necessary physic after such a debauch of sweet stuff was to come later, when they were surfeited.

In extenuation of his subsequent career, it should be remembered that De Soto suffered from a blind spot in his mental vision. Like many men of great talents he at first had mistakenly believed that the world sincerely wished to better itself. If so, why shouldn't it be eager to reach the best possible state in one quick stride, instead of blundering this way and that like a drunken imbecile, and getting nowhere in a thousand years? De Soto here made the usual mistake of the super-intelligent in thinking that his own clear vision would satisfy the blind.

The first six months at the Foundation passed like a Persian dream before the half closed eyes of the purring trustees. Without the least suspicion that their brilliant young employee was feeding them all this unnecessary wealth for purposes of his own, they squatted like drowsy bullfrogs on a warm summer day in their golden swamp, expanding their already enormous business and swelling to the bursting point with financial pride. The fatter they grew the faster they bloated. But De Soto, like the true artist he was, deferred the adroit pin prick which would deflate them all, until power had become a fixed habit with them and inordinate expenditure their means of keeping alive.

One example of De Soto's methods will suffice. It is already a classic the world over, but its retelling here may throw some light on his general campaign, which even now is not well understood outside of a very narrow circle. His first great financial triumph was a mere byproduct of his toy transformer and storage battery—the project which got him his appointment to the staff of the Foundation. He had undertaken to emprison twenty million volts in a small box, and to control his trapped devil in any way the trustees desired. In short, he promised to put the elaborate twenty million-volt laboratory, and all of its rivals, completely out of business, and to do it for an expenditure not exceeding five thousand dollars. When the trustees remembered that their high tension laboratory had cost close on three million dollars, they saw the most obvious commercial possibilities in a flash. Although there was as yet no practical use for such a devil box as De Soto promised to deliver—unless the military and naval authorities might be tempted to flirt with it—the trustees had faith enough in pure science to believe that somehow, some day, the dollars would gush out of that evil box. Some young man as brilliant in a practical way as De Soto seemed to be scientifically would come along like Moses with the right kind of stick in his hand. Then, with one resounding smack, he would smite the useless black box, crack it wide open, and let the golden deluge drown the trustees in dividends.

De Soto did not wait for a greater than he to enrich the Foundation beyond its thirstiest dreams. He did it himself, almost in his sleep. One detail must be settled before the box itself could be constructed. This was a revision of the whole theory and practice of insulation. The huge strings of earthenware mushrooms that made the long distance transmission of high voltage possible obviously would not do. The high tension lines from the mountains to the cities carried but a paltry two or three hundred thousand volts; De Soto must handle twenty million. To insulate against such a pressure with glazed earthenware, or with any of the known substitutes, would require a mass of dead material equivalent to several hundred times the small box into which De Soto planned to compress his entire apparatus, insulation, transformers and all.

Following a hint he had absorbed in his exhaustive reading, he saw that the true way out of the difficulty was not the building of more and more massive resistances of earthenware and the rest, but the practical construction of material films thinner than the most tenuous soap bubbles. These must be manufactured cheaply and deposited directly on the wire carrying the high current as an invisible sheath not over a few atoms in thickness—the thinner the better. With Crane's help, De Soto had perfected the working drawings and specifications of the process three days after he joined the Foundation. The plans were turned over to the technical staff for practical development, and in two weeks the Foundation had staked out its first El Dorado.

To their surprise, the trustees discovered that De Soto was an adept in the finesse of service as understood by them. It was his campaign that they launched against their innocent competitors. The Klickitat Lake Municipal Power Company, having just completed its giant power plants in the Cascades, was calling for bids on the insulation of its three hundred thousand-volt trunk line. De Soto saw the Foundation's great opportunity and, incidentally, his own. The Power Company belonged to the people of Seattle. It was a public enterprise, supported entirely by taxes. By eliminating dividends to stockholders, the public hoped to obtain its power and light at a cost much below the current rates. Why not, De Soto suggested to the trustees, donate the required insulation to the public? The trustees saw the light and smothered their indecorous grins. By presenting the public with a few thousand dollars worth of the new insulated wire, and saving the oppressed taxpayers several hundred thousand, the Foundation would net an incalculable amount of free advertising and the good will of the people.

The engineers of the Municipal Power Company came, saw a four days' demonstration, and were conquered. High steel towers, tons of insulation, and expensive copper cables were all replaceable by a thin wire sheathed in the new film and suspended from trees, telephone poles or broomsticks as convenient. In the words of their chief, it was a knock-out. It was. In five weeks the Erickson Foundation had a monopoly on insulation the world over, and one great corporation after another, from San Francisco to New York, from Manchester to Brussels, went flat.

This was but the beginning. De Soto, with Mephistophelean ingenuity, talked the not ungenerous trustees into trebling the price of the new, simple insulation the moment their strongest competitor collapsed. Having created a new necessity of modern life, the Foundation had the electrical industry at its mercy. To their credit it should be recorded that the trustees did not yield without a short struggle to De Soto's cynical importunities. Their profits already were outrageous; why make them sheerly indecent? De Soto could have enlightened them in one sentence, had he felt inclined. But the time for the dazzling revelations of the surpassing splendor which was to burst upon the trustees was not yet ripe; first they must be educated. They could not withstand this frank young man's magnetic charm. One and all they agreed that he was irresistible. He was.

All that De Soto asked in addition to his modest salary and rapidly mounting royalties was the time occasionally to undertake a piece of pure, unpractical research. As these short excursions into science for its own sake always resulted in some radical improvement of existing luxuries that sent whole businesses to the wall, the trustees humored him. There was the little matter of high vacua, for example. De Soto begged for a ten days' holiday in which he began his explorations of the hardest cosmic rays. First he must obtain a practically perfect vacuum. The so-called vacua of hard X-ray technique, where billions of molecules of gas remain in each cubic inch after the diffusion pump has done its utmost, were of no use in his project. He needed a tube from which all but a hundred or two of the ultimate particles of matter have been withdrawn. Again Crane and other members of the staff helped him with the mechanical details, and again he triumphed completely.

As a byproduct he revolutionized the industry of making electric light globes and radio tubes, cutting down the cost of exhausting these to a fraction of a per cent of what it had been. The trustees beamed on him, and told him to take a year's vacation if he wished. With a subtle smile, which they failed to interpret correctly, he refused. Later, he said, he would take a real holiday. They thought he was merely modest in a decent, humble way, like all good scientists—of their rather uninstructed imaginations. De Soto had a withering contempt for science and all its works as evidenced by the age in which he was condemned to live and be bored. But his frank geniality would have blinded almost anyone to the smouldering volcano which it concealed.

In addition to the commercial byproducts of his earlier genius, another, of a purely social character, was to have far-reaching consequences for himself. Partly to spite Crane, and partly because he had no insight into the more morbid aspects of human character, Kent insisted that De Soto occupy the guest suite in his house indefinitely. De Soto consented, chiefly because he disliked the bother of hunting up suitable quarters for himself. He breakfasted with Kent and Alice, but took his other meals out, except for an occasional festive dinner in honor of some new triumph of his commercialized genius.

The inevitable happened. Alice fell hopelessly, degradingly in love with him. Before the irresistible charm of his resonant voice, his perpetual high spirits—they seemed high to her—and his vibrant vitality, she abased herself utterly. His careless words of greeting became her treasured pearls of seraphic wisdom and celestial love. She is not to be unduly censured for her blindness; De Soto might have had any girl he fancied for the trouble of asking. But he never bothered to ask. He saw what had happened to Alice and it did not even amuse him. Nor did it move him to intercede in behalf of her father when the lightning struck him.

The President of the Board of Trustees was blessed with a long and accurate memory. His spite against Crane evaporated and condensed on the hapless director. Kent, he remembered, had proposed the plan for forcing Crane to resign. It had resulted in making the President feel foolish—a disagreeable sensation to any self-respecting man. Accordingly, when De Soto began conquering the electrical industry, the President decided that Kent was no longer necessary to the prosperity of the Erickson Foundation.

"How would you like to be director?" he jovially inquired of De Soto on the morning of exactly the hundred and eightieth day of De Soto's contract.

"How about Mr. Kent? What will he do?"

"Go fishing," the President hazarded with a slow smile of doubtful sincerity. "You see eye to eye with me in this matter," he continued, and De Soto did not deny the allegation. "Mr. Kent is no longer necessary to us. What do the trumpery eight or ten millions a year that he begs from old paupers amount to, anyway? The royalties from your new oil switch alone—the cheapest thing you've done—make all that Kent brings in look like a Mexican dollar. You take hold of things and show the world what a real, up-to-date business administration is. Accept now. There's a good fellow."

De Soto lazily stretched his arms and yawned.

"All right. I'll quadruple your profits in a week."

The President was about to shake the new director warmly by the hand when the latter, for no apparent reason, doubled up in an uncontrollable spasm of laughter. Thinking De Soto was enjoying the joke on Kent, the President joined in the whole-souled shout. The harder he laughed the worse De Soto became. At last, after a severe tussle, the swarthy young man gained control of himself and stood gazing with humid eyes at the President.

"This is rich," he gasped. "You will tell him of course?"

The President nodded, and De Soto went soberly to the workshops to supervise the construction of the last unit of his cosmic ray generator. That afternoon Kent broke the news to Alice.

"I have saved practically nothing," he confessed bitterly. "Well, I can peddle life insurance till some place offers me a position. We shall have to vacate this residence within a month. It will be De Soto's now. He's director; I'm down and out."

"Perhaps he will ask us to stay here until we get settled," Alice suggested, a sinking at her heart.

"Impossible! To accept hospitality from a man who has stabbed me in the back? Never!"

But he did, that very evening, when De Soto, in response to a humble hint from Alice, indifferently invited Kent and his daughter to stay as long as they liked. They were to manage the house; he would pay the expenses. Things were to go on precisely as before. De Soto did not care whether they stayed or went, and the considerable expense of keeping up the establishment would not make even a dent in his weekly royalties. Should the Kents finally decide to leave, he would have a housekeeper. All he asked was that he be spared the bother of settling down to a new regime.

The morning after the disaster, Alice rose much earlier than usual and waylaid the generous protector of the poor before he entered the breakfast room. The utter self-abasement of her thanks seemed to rouse De Soto's smouldering contempt.

"Are you a human being?" he demanded roughly.

She failed to comprehend the blasting sarcasm of his brutal question.

"Of course," she laughed.

"Well, then—" he began and stopped abruptly. A brilliant

idea for an experiment had just flashed into his mind. His harsh tone softened, and he laughed in the mellow way that he knew was music to the deluded girl's soul.

"What I was going to remark," he continued, "was simply this. You are human; so am I. All that I have done for you is nothing. Nothing!" he repeated with furious emphasis. "If human beings can't do so little as nothing for one another, and not have to be thanked for it, they are no better than hogs. Or," he added after a reflective pause, "than a certain cook and waiter I saw about six months ago. So please never refer to this matter again. Stay here as long as you and your father wish and let things go on exactly as they did before. I'm comfortable; why shouldn't you be?"

His generous words, whose acid sting she missed completely, turned her blind love to dumb adoration. She was his whenever he wanted her. But he did not want her—yet. First he must perfect his generator.

All through the spectacular months of De Soto's rocket rise to world fame, Crane served his superior as faithfully as he could in the tasks that fell his way. Without the slightest twinge of jealousy, professional or personal, Crane admitted that De Soto's mind soared above his own a universe away. The young man never seemed to think out his problems or to reason painfully from one verified guess to the next, as even the greatest scientists do, except in their two or three flashes of blinding genius. The beginning and the end were alike to him; the beginning was the wish to accomplish some bold project, the end its accomplishment. De Soto's method was like a continuous streak of lightning. For all that Crane could see, nature offered no puzzle more perplexing to De Soto's easy skill than breathing is to a normal man. Thus far Crane was one with the rest of the staff. Where he parted company with them was deep down in his secret thoughts.

Crane's first impression of De Soto remained as vivid as

ever. Outwardly the two men were on the friendliest terms. What the young king of all inventors thought of his technical assistant he kept to himself; Crane's opinion of his chief was too dangerous to be shared with the rest of the staff. Brown was his one confidant. Their common nightmare with the black widows in the laboratory had established a bond between the two that nothing could break. They seldom referred to their unnatural adventure, but both knew that it was at the bottom of their unreserved friendship. Nor did Crane ever allude to the great discovery which the doctor announced that he had made with his microscope.

On the evening of Kent's dismissal from the Erickson Foundation, the doctor dropped in to spend an hour or two with Crane.

"Wilkes called me up this afternoon," he began, when Crane had made him comfortable.

"Wilkes? Oh, yes, I remember. He's in biology over at the university, isn't he?"

"Up to his neck in it. I doubt if he's deeper in than that. Did I ever tell you about my little spat with him six months ago?" Crane shook his head, and the doctor freely confessed the humiliating episode of the vanishing protozoa. "Wilkes thought I was crazy drunk that morning. He told me this afternoon that he poured that priceless bottle of your historic bathwater down the kitchen sink the moment I was out of his house. He had promised me like a gentleman and a scholar that he would get one of the chemists to analyze it. When I asked for a report some ten days later, he assured me the chemist had found nothing but pure water with the usual traces of organic matter and minerals—lime, and such stuff—that are in all tapwater. Like a fool I believed him. Now he's kicking himself for the scurvy trick he played me."

"Why?" Crane demanded, scenting a clue at last to the incomprehensible mystery of the black widows.

"It's a long story. I'll cut it short. You remember that poisoned fish scare we had six months ago?"

"When the fish in the bay turned up all spotted, and you asked me to sterilize my bathwater?"

"Yes. And you remember how it passed off in a day or two? The fish seemed to recover completely, or else all the affected ones died. Anyhow, no biologist in this part of the world had curiosity enough to ask the Health Department for one of the spotted fish to examine. I don't blame them—I didn't think of it myself. Now here comes the fortunate part. Some crooked inspector in the department, instead of destroying the fish seized in the markets as required by law, sold the lot to a Japanese cannery down the coast. Some of that canned fish found its way to the table of Professor Hayashi, the expert in parasitology at the Technical College in Tokyo. The discolorations on the skin caught his eye at once. To cut a long story short, Hayashi's microscopical examination of the diseased skin revealed myriads of protozoa—the most rudimentary forms of animals—of species totally new to science. Being a German-trained Japanese expert on parasites, Hayashi went without food or sleep until he had prepared an exhaustive series of microphotographs of these strange new beasts."

The doctor paused long enough in his story to extract half a dozen beautiful photographs and the same number of his own hasty sketches, made the night when he examined Crane's bathwater, from his pocketbook.

"Compare the photographs with the sketches. The photographs are Hayashi's, the sketches mine. Professor Wilkes gave me those specimens of Hayashi's work this afternoon. It seems that Hayashi picked on Wilkes as being the likeliest man to recognize the protozoa if they were known. The fish had been canned a few miles south of here, according to the label on the can. Hence, Wilkes, being professor at the university here, and presumably not dead, would know all about what lay at his own back door. Unfortunately, Wilkes had poured the most conclusive evidence, and my one chance to be famous with it, down the kitchen sink. He called me up this afternoon to ask what is to be done about it. I have several suggestions, but I'm not going to share them with him. He's too fond of his kitchen. Well, what do you think of my sketches? As true to life as the microphotographs, aren't they?"

"You copied Hayashi's?"

The doctor laughed. "No, I drew them from what I saw under the microscope in that bathwater I got from you six months ago. It's a great pity that you recovered as quickly as you did from that unique itch."

"I may be able to oblige you again, in a day or two," Crane grinned. "But go on. You were going to say something."

"Only this. I fell for the biggest fool on earth. By the margin of a single stupid mistake, I have lost a new universe. If I had not gone over to see old Wilkes that morning, I should have kept on believing in my work. Now Hayashi will get whatever credit there may be in it. This stuff is brand new, I tell you!' he exclaimed, warming to his beloved protozoa. "These things could never have evolved from anything we know. Structurally they are entirely different from any that have ever been described. And to think that I saw them living and multiplying under the lenses of my own microscope!"

The doctor lapsed into moody silence. "Well," he concluded, "it's too late now. Still, Hayashi doesn't know the whole story. Until he or someone else explains why those prolific little pests stopped multiplying in the sea, we haven't even begun to explain them. I should have expected every fish in the Pacific Ocean to be as gorgeous as a rainbow two months after the infection started in our harbor. But it stopped suddenly and absolutely in a day. Why?"

"Ask me another. I'm not good at riddles. Want to see some more blood-red bathwater?"

"Where?" the doctor exclaimed, leaping to his feet.

"In the bathtub, of course."

"Lead me to it!"

"It isn't there yet. But, if De Soto and I have any luck tomorrow, I'll brew you twenty gallons of the reddest water you ever saw. He and I have about finished his first two million-volt X-ray tube. It's no longer than my forearm. Built on entirely new principles. So, if my exposure to the hard X-rays had anything to do with the infected state of my skin, we should know it by tomorrow night. What do you think?"

Brown looked depressed. There were so many factors that his optimistic friend had overlooked that the doctor dared not feel enthusiastic.

"You forget the time element," he said. "One exposure, even of thirty hours, may not be sufficient by itself. Why did those fish suddenly recover? No; there is something beyond hard X-rays at work in all this. Your luck six months ago may have been only an accident due to a concurrence of causes that won't happen together again in a million years."

"Cheer up. We can only try, you know. This time tomorrow I may be wishing I were dead."

"I hope so," the doctor replied absently, dreaming of his lost universe. "In case an accident does happen, we must be prepared. You say De Soto's tube is built on entirely new principles?"

"New from beginning to end, from anode to cathode—like everything else he does. Lord! I wish I had one per cent of the brains that kid has."

"Then you might never duplicate that scarlet bath. It was the most brilliant thing you ever did. The very blunders of your own two-million-volt tube may have been responsible for what happened."

"Possibly," Crane admitted. "X-rays were discovered half by accident. It begins to look as if my precious tube may have been in the same class. Something that Bork and I did in spite of ourselves touched off a real discovery—which we succeeded in smothering between us."

"And I, too," Brown sighed, "missed the essential thing. Well, we must be prepared. Will you take half a dozen of Bertha's eggs with you to the laboratory tomorrow? Don't let De Soto know about them. Keep them wrapped up in the pockets of your working coat."

"Don't you think," Crane suggested, "it is about time for you to give me a hint of your theory? I won't steal it."

"You'd be insane if you did. Even I haven't the nerve to talk it over with a biologist. Very well, here goes—your itch, my protozoa, and our black widows."

For an hour and a half the doctor defended the shrewd guesses and bold theories which he had devised to account for the apparently unnatural adventures in which he and Crane had participated. To Crane's frequent interruptions that the physics of the explanation, at least, was too wild for any sane man to listen to, the doctor retorted that only a hopeless pair of lunatics could have witnessed what Crane and he saw with their own eyes in the twenty-million-volt laboratory. Was that a fact of experience, or was it not? Did they see black widows by the hundred as huge as spider crabs, or didn't they? Well, then, the doctor continued somewhat irrationally, if nature can upset her so-called facts to suit herself, why can't she equally well break the puerile laws which we imagine for her discomfort? Hadn't all of the great generalizations of Nineteenth Century physics been scrapped or changed out of all recognition in the first three decades of the Twentieth?

Yes, Crane admitted, but why try to answer the old insanities with new ones even more insane? To which Brown replied that when in a lunatic asylum—meaning modern physics

—do as the lunatics do; namely, cut your theories according to your facts.

It was a wild argument and a merry one. The climax was a roar of laughter from both disputants simultaneously. For it occurred to them that they were but slightly parodying the proceedings of two recent world scientific congresses which they had attended—Crane in physics, Brown in biology. No layman could have detected the slightest difference between their fantastic arguments and the profound debates that will make those great congresses forever memorable in the history of science.

"That settles it," Crane gasped, wiping the tears from his eyes. "I'll take a whole crate of eggs to the laboratory tomorrow."

"Don't!" the doctor implored. "Half a dozen, no more. I don't want a man as brainy as young De Soto is to suspect what I'm up to. The least hint to a man of his intelligence, and I'm dished. He would go into it for all he's worth and clean it up from beginning to end in a week."

"Perhaps he has already," Crane suggested quizzically. "By the way, all our crazy talk made me forget my bit of real news. Kent's fired."

"What! When?"

"This morning. De Soto told me all about it. The trustees, it seems, decided they didn't need Kent any longer, now that De Soto is bringing in the money by the trainload. So they kicked him out."

"The low hounds!"

"Oh, I don't know. Business is business. They made De Soto director."

"And he accepted?"

"Why not? They had no further use for Kent. Live and learn ethics, doctor. All professions aren't like yours. I only wish they were. Even De Soto seemed disgusted with humanity in general and with the President in particular. It was the first time I have seen him show any human feeling. He was quite glum all the afternoon till about six o'clock, when we finally managed to put his new tube together. Then he yawned—he's always yawning—and began laughing like the devil. I mean it; he laughed exactly as a good fundamentalist thinks the devil laughs when he sees some nice boy downing his first drink of whiskey. It actually made my flesh creep. I don't like that young fellow."

"You still distrust him?"

"Yes, and I don't know why. Sometimes I have a feeling that he is about five million years old."

"Absurd. You had better be going to bed, and so had I. Don't forget to call for those six eggs on your way to the laboratory tomorrow."

"I won't. How's Bertha, anyway?"

"Fine. She and Roderick have done nobly. My back garden is full of broilers now, and I haven't the heart to eat one of them. The housekeeper threatens to give notice unless I sell a dozen. She can go if she likes. Bertha is a joy, and I wouldn't hurt her feelings to pacify fifty housekeepers. Well, I'm going. See you in the morning."

Crane duly called for the six new laid eggs the next morning and joined De Soto at the laboratory at eight o'clock sharp. For once in his life De Soto seemed to be laboring under the strain of repressed excitement. Usually he was somewhat indifferent about his brilliant work, not to say bored by it all; today he could scarcely wait for Crane to begin trying out the new tube.

"Put on these," he ordered, handing Crane a long shroud of crackly transparent material, with overshoes, gloves and hood of the same. He himself was already armored for their dangerous work. The crackly stuff, not unlike the thinnest isinglass, was another by-product of De Soto's incessant inventiveness. It had grown out of the work on insulators, although dependent upon different principles, and was sufficient to block completely all rays hard enough to penetrate forty feet of solid lead. Against the

X-rays Crane expected to generate in the new tube it was more than ample protection.

The walls, floor, ceiling and window panes of De Soto's small private laboratory were closely "papered" with this thin, transparent ray insulation. It would not do to have stray radiations penetrating into adjacent laboratories and deranging delicate electrical apparatus.

Crane waved the proffered garments aside.

"No, thanks. I've worked around rays as hard as these, and I'm still kicking."

"Better not try it again," De Soto advised, with just the suspicion of a threat in his voice. "Put them on."

"Sorry, but I must decline," Crane replied with a defiant grin, his square jaw thrust slightly forward. "You see, I want to do a little experimenting myself."

For three seconds that seemed to Crane to stretch to three eternities, De Soto's blazing black eyes fixed upon his. "What is happening to me?" Crane thought. "I feel as if my brain were being torn to pieces."

"What experiment are you going to do?" he heard De Soto's voice asking in tones of deadly calm.

"Nothing much," Crane replied. He felt sane again. "I just wanted to verify my guess that rays as hard as these cannot affect human cells."

"Human cells?" the deadly voice echoed, slightly emphasizing the first word.

"Yes—the units from which our muscles, bones and nerves are built up. You know what I mean."

"I know what human cells are," De Soto said slowly, with deliberate ambiguity. His tone implied that he suspected Crane of an interest in cells not human. "Put on these things, and let us get to work."

"I have told you I prefer not to."

De Soto's eyes flashed ominously.

"Don't make it my unpleasant duty to discharge you for insubordination as my first action as director."

Crane hesitated. For a second he was tempted to defy the new director and take his medicine. Then he remembered why he had swallowed his pride in the first place when Kent tried to make him resign. He also thought of Brown, and the disappointment of his friend should he fail in the matter of the eggs. Without a word he shed his working coat and hung it on a hook behind the door. De Soto followed him with his burning eyes.

"Why did you take off your coat? This material weighs very little. You won't be too warm with your coat on under this. I'm wearing mine."

"I'll feel freer without so many clothes," Crane replied offhandedly.

"You have something in the pockets of your coat that you wish to be exposed to the full effect of these rays. Take out whatever you have, and destroy it. There's an oxy-acetylene torch by that bench."

Crane tried to bluff it out. Going to his coat, he extracted the half dozen new laid eggs and casually exhibited them for De Soto's inspection.

"My lunch," he explained.

Without replying, De Soto took one of the eggs and held it up to the light.

"You eat them raw?"

"Are they raw?" Crane asked in well feigned astonishment. "That girl at the lunch counter must have done it as a joke on me—or else she was rushed and made a mistake."

For answer De Soto took the six eggs, one at a time and smashed them against the wall behind Crane.

"I'll take you out to lunch," he laughed good-humoredly. Suddenly his whole manner darkened. His eyes blazing, he shot an accusing question at the pale face before him.

"Did you come back here last night after I left?"

"I don't get your drift."

"Do you know what kind of a tube this is?"

"A two-million-volt X-ray, if it's the one I've been helping you build."

"It is the same one. Could you make another like it?"

"How could I? You made the cathode and the anode yourself. I don't know the first thing about their construction. And I have no idea what that thing like a triple grid in the middle is for. All I know about your tube is what you've told me. You said it was for hard X-rays. It's all new to me."

"It might be a device of great commercial value?"

"For all I know it might."

"Go to the office and get your time. You are discharged."

Crane turned on his heel and walked out without a word.

7

WARNED

CRANE DID NOT BOTHER ABOUT HIS PAY CHECK at once. It could be collected later. For the moment a matter of greater importance pressed. He sauntered into the small chemical laboratory where tests of materials were carried out in connection with the electrical work. Only one man, a technical assistant, was in the laboratory. Looking up from his work he greeted Crane with a curt nod. Like the rest of the staff he treated Crane coldly since the latter's degradation in rank.

"I'm coloring a meerschaum pipe," Crane volunteered. "Got any beeswax or anything of that kind?"

"There's some in the drawer under the bench—there by the window."

"Thanks," Crane helped himself to four cakes and walked out. On reaching his own office, he locked the door, and proceeded to take impressions of all his keys to the laboratories and workshops of the Foundation. It would be a simple matter to have duplicates made at some obscure shop in another quarter of the city. "They may put me in jail before I finish," he grinned to himself. "Anyhow, I'll give them a run for their money first."

In the business office Crane explained that he had come to collect his pay.

"But it is only the fifteenth of the month, Doctor Crane," the clerk objected. "Of course, if you want an advance, I dare-say it can be arranged. I'll have to ask the head bookkeeper."

"Don't bother. I'm fired."

"Fired? What for?"

"Being too smart for our new director. Make out the usual month's bonus for discharge without notice."

"Sorry, but I can't. We don't give any bonus now."

"Since when?"

"The new rule went in yesterday, before Mr. Kent was discharged. The trustees made the rule before they elected Mr. De Soto."

"I see. They don't overlook anything, do they? Is the President of the Board anywhere about?" The clerk nodded. "All right, call him out, will you? It's the last favor I shall ever ask of anyone connected with the Erickson. Here are my keys."

"But I can't do that. He's busy."

"Never mind. Tell him I'm here with an urgent message from Mr. De Soto. Honest; I'm not fooling."

The clerk was fooled. Presently the high potentate himself hurried out of his lair to receive the urgent message in person.

"Yes, Doctor Crane? Perhaps you had better come into my office."

"Perhaps I had."

"Now, what is it?" the President demanded when the door was closed. Crane looked him squarely in the eyes.

"You can go to hell!" he said.

"I shall ask the Director to discharge you," the President roared when he recovered his breath.

"Too late. He just did it. I meant that message from myself to you personally. You're a pretty cheap sort of skate. Nevertheless, I'm rather sorry for you and the rest of this firm of pawnbrokers. De Soto is making you all multimillionaires in record time, isn't he? And he loves you all better than if you were his brothers? Fine. Take it from me, he hates every last one of you worse than a rat hates rat-poison. I know that young fellow; you don't. Look out that he doesn't leave you flat. That's all."

The President's face went a pasty yellow. For the first time it dawned on him that De Soto might have been laughing at him, not at the unfortunate Kent, when the latter was so swiftly fired. That laugh, in retrospect, had a peculiar quality. Swallowing what remained of his pride, the President motioned to Crane to be seated. Crane declined, and the President affected not to have noticed the rebuff.

"Of course, I should be very angry with you," he began jovially, "for what you said when you first came in. However, boys will be boys, eh? Now, don't get sore because you think you've lost your job. Perhaps you haven't. How would you like to be my technical secretary? I get hundreds of letters from all sorts of people that I can't answer properly. You know how it is in my position. What about it? Fifty per cent increase in salary, of course."

Crane slowly shook his head. Business was business, as no one understood better than he. The President's purpose was evident. But Crane was not to be bribed into the dangerous job of spying on De Soto while an employee of the Foundation.

"I'm afraid it wouldn't do," he said. "The Foundation can't afford to antagonize its new director."

The President regarded him long and thoughtfully before replying.

"You're loyal to us, aren't you?" he remarked with evident sincerity, as if the discovery surprised him—as indeed it well might.

"I have some pride in my profession, if that is what you mean," Crane admitted.

"Am I to infer that you think Mr. De Soto lacks professional pride?"

"De Soto is too brainy to take pride in anything. That's the trouble with him."

"H'm. You think he may be playing a game of his own?" Crane nodded slightly. "What sort of a game?"

"As De Soto has several thousand times the mind that I have, I can't guess what his game is. It may all be my imagination."

The President paced the carpet in silence. Coming to a halt before Crane, he went to the roots of his doubt.

"Why are you telling me all this? Can't you see that it may be to your own disadvantage?"

"That's easy. Because I hate De Soto. I know," he continued with a dry smile, "that men don't hate one another nowadays outside of the movies. It simply isn't done—in the way that I hate De Soto. Still, I do, and that is the plain fact."

"Why do you hate him?"

"How can I tell? It may be repressed professional jealousy, for all I know. More probably it is based on fear—or cowardice, if you like to put it so. I'm afraid of what he may do to me, and to the Foundation."

"What do you suspect?"

"Nothing definite. Everything about the man, except his scientific ability, is vaguely rotten. My guess is that he is planning something brand new to take us all by surprise. We probably shan't know what has happened to us until it is all over."

"Would you care to act as my agent to keep an eye on things? No one need know that you are connected with me or with the Foundation in any way."

"I'm not a detective. Still, in case I get into trouble I may as well tell you that I intend to keep an eye on Mr. De Soto on my own account."

The President touched a button. When the clerk appeared he requested him to make out a check to Doctor Crane for the amount of five years' salary, "as a token of appreciation," he added for the clerk's misinformation, "for his excellent services to the Foundation in the past." Crane did not object; he felt that he might need the money before settling with De Soto. It was a worthy cause.

"Now, Doctor Crane, perhaps you can be more explicit. What do you plan?"

"Just what I have told you. If I were rich," he added with a grin, "I should go right back to the laboratory now and shoot De Soto. Then I'd hire the best lawyer in the country to get me off. Your new Director is more dangerous than any mad dog in the country."

"In what way dangerous? You are making pretty serious charges, you know."

"I can't tell you definitely, because I don't understand myself. But I can convince you, or any other business man, that De Soto had better be handled with care. The Erickson has sent quite a few businesses to the wall recently, hasn't it? The whole industry of insulation as it was a short time ago, for example. We all saw how it was done. Has it never occurred to the trustees that the Erickson might go the same way?"

The point was obvious, and the President saw it. A clammy perspiration prickled out on various parts of his body. He felt quite ill.

"But Mr. De Soto is under contract to us for ten years," he protested weakly.

"What of it? He won't have to break his contract to break you."

"He can't be so unscrupulous as you suggest. Who ever heard of a scientist turning crook like that? All the men on this staff are as honest as the day is long."

"Perhaps it is the other way about. De Soto may have been a crook before he took up science. According to his story he comes from Buenos Aires. I've pumped him. He never saw South America. As to contracts, he has as little respect for them as he has for—I don't know what. Why shouldn't he leave you in the lurch tomorrow, if he likes, and go over to some of your competitors? Before the courts had settled the row, you would be flat broke."

The President was now perspiring freely. If De Soto could lie about his native country, why not about business matters? The possibilities of a broken contract were too obvious and too awful to be contemplated in silence.

"What would you do in our case—provided your suspicions are justified?"

"Sell out to my nearest competitors. Let them absorb the shock. It's coming."

"We can't," the President almost groaned. "Business details —I needn't bother you. But we can't." He tried to believe that Crane was merely letting his imagination run wild, but he could not. Innumerable slight inconsistencies of word and action on De Soto's part loomed up now with a sinister significance. For the first time the President suspected that De Soto's perpetual good humor and high spirits were the rather cheap disguise assumed by a man who had much to conceal. "I'm glad you have warned us," he admitted unhappily. "If you learn anything you will let us know? You won't find us ungrateful. Business is business, you know," he concluded with a rueful attempt at jocularity.

"I know," Crane retorted grimly. "So does Kent. That's why I came to tell you what I did first. I had no idea it would end this way. Thanks for the check; I'll need it. And please remember that if I get into hot water, it will pay the Foundation handsomely to fish me out. We are working on the same job, but for different reasons. Good morning, and thanks again."

They parted almost on good terms, but not quite. Crane still despised the Foundation's business methods; the President resented Crane's greater penetration in seeing a practical danger which he himself should have noticed months ago.

Crane's guess as to De Soto's probable financial tactics was shrewd but wrong. De Soto had no intention of using any of the obvious devices imagined by Crane. They lacked humor, and De Soto enjoyed nothing so much as a good laugh.

The next, and last, person to be warned was Alice. She, presumably, would still be living at the Director's residence; De Soto could not have turned the Kents into the street already. Hailing a cab, Crane directed the driver to take him to a locksmith's on the other side of the city. In the dingy little shop he asked the dried-up old tinker to duplicate the keys impressed in the wax, and to have them finished by five o'clock. "A rush order," he explained. "My partner lost his office keys, and I lent him mine." Crane then drove to the Director's residence and asked for Mr. Kent. The man who answered replied that Mr. Kent was out.

"Is Miss Kent at home?"

"I will inquire."

Before surrendering his card, Crane scribbled on it, "May I see you for a few minutes? Important."

Although he had not seen Alice since the night when she passed him in her car with the newly hired De Soto, he felt reasonably certain of her state of mind. Indeed Kent had spared no hint to make it plain that Alice, once fond of Crane, had now no further use for him, and that he need not trouble to call. While waiting for her to come down, Crane briefly reviewed his own feelings toward her, in order to be sure of himself and not blunder in his delicate task. Had he ever loved her?

Looking back on their friendship he admitted that he might have loved her, if circumstances had permitted their bantering goodfellowship to ripen, but that he had not. Her sudden and complete discarding of his friendship argued that she also had never cared seriously for him. If her indifference had been a pose to quicken his love for her, she would not have let it drag on indefinitely as she had. Her infatuation for De Soto, like his for her—according to Kent's sly, optimistic hints dropped to exasperate Crane—was genuine. Finally, Crane admitted to himself, he was a bachelor by instinct, as he had told Bork that afternoon in the high tension laboratory. A family was, after all, to a man of his temperament, only a lot of grief, as he had declared. He much preferred to live his own life, with its long working hours, its snatches of sleep, irregular meals, and scientific fights with Brown, to all the comforts of any home.

Alice entered, pale and distraught. He saw that she was still beautiful. But she had aged ten years, and all her happy spontaneity was gone. Worry over her father's plight, he speculated, could not account for all of the sad change. And instantly his hatred against De Soto doubled. He himself had never loved her, he now realized fully. But she had been such a good fellow that he resented De Soto's malign influence over her as fiercely as if she were his wife.

"It must be months since I've seen you, Doctor Crane," she said, extending her hand.

"Several, Miss Kent," he replied, noticing the formality of her address. Well, he could dance to any tune she called.

"You wished to see me about something?"

"Yes." He went to the point at once. "About Mr. De Soto."

"I must refuse to discuss Mr. De Soto with you," she interrupted hastily, her cheeks flaming.

"I don't intend to discuss him. As one human being to another, I shall tell you a fact that you should know. I do not apologize for what I say. It is none of my business. That is true. And it violates every decency of good society. What of it?"

"I won't listen!" she cried, putting her hands over her ears, and starting for the door. "Please go."

In one stride he overtook her. Forcing her hands to her sides, he delivered his message.

"De Soto is rotten to the core. He is not fit for any decent human being to associate with. If you marry him, you will kill yourself to be rid of him. Use your eyes and your brains!"

He released her hands and she fled, sobbing.

"Well, I've done it," he muttered. "It will make her watch him anyway. But I'm too late. That fool Kent!"

From the Kent's home he hurried to call on Brown. The doctor was in his office busy with a patient. At last the sufferer left, and Crane was admitted.

"Hullo," Brown exclaimed, "Not working today? Don't say you're itching again," he cried, hopefully.

"No such luck," Crane grinned. "But I'm hot enough. De Soto fired me the first thing this morning."

In answer to the doctor's solicitous questions, Crane briefly told the whole story, including his interviews with the President and Alice. Brown was shocked. The thought that his half dozen eggs had brought his friend to grief filled him with remorse and dismay.

"What will you do now?" he asked.

"Take things easy for a time. Till five o'clock this afternoon, to be exact. I forgot to mention that I'm having duplicate keys made to all the laboratories and workshops of the Erickson. They'll be done at five."

"What! You don't mean to say you're going to burglarize the place? That is what it would amount to, now that you are discharged. Better not try anything so foolish."

"It may be foolish, but I'm set on it. And I'm going to make my first attempt about one o'clock tomorrow morning—just when it's darkest. I know when the watchman makes his rounds to the various buildings. He and I will contrive not to meet. There's no danger worth mentioning. If they catch me at it, I shall appeal to the President and ask him to lie me out of the scrape. But I'm going to find out what De Soto is up to, no matter what it costs. The sooner the better."

"I agree to the last," Brown seconded. "The way he smashed those eggs when he found they were raw looks bad. I don't half like it. De Soto knows something he shouldn't. We've got to learn what it is."

"We?" Crane echoed.

"Yes. You and I. Now, don't argue. But for me, you never would have got into this mess. I'm going with you to stand guard and see that you don't go to the penitentiary. Don't put too much faith in your friend the President."

"I don't," Crane grinned. "If he saw a chance of making a dollar out of it, he would double-cross himself. Well, I shan't mind if you do come. Two will be safer than one. You can slip away if anything unpleasant happens. Shall I call for you about twelve tonight?"

"All right. It has just occurred to me that we shall have a glorious chance to perform a crucial experiment if we can get hold of De Soto's tube for a second or two. Can I bring Bertha? I'll drug her before we start so she won't squawk and give us away."

"Bring her along. She can stand the risk, if we can."

De Soto's morning, after he had discharged Crane, passed pleasantly enough. The lack of an assistant did not inconvenience him, as he was essentially a lonely worker. In fact he had retained Crane more as a blind to the trustees than as a help to himself. By acquiescing in what seemed to be their desires with regard to Crane, he not only proved himself a good cooperator, but also a decent, human fellow, whose work need not be carried out in secret. Actually he feared no ordinary physicist. To understand what De Soto was doing, a spy would require at least his own intelligence.

When Crane walked out, De Soto's first act was to lock the door and pull down all the window shades. He was now secure from uninvited observation. No ray could penetrate the insulated walls, floor, windows and ceiling to give workers in neighboring laboratories a clue to the nature of the radiations which De Soto hoped to generate.

He was about to set his tube in operation when he paused thoughtfully, as if in doubt. Going to the closet, where the suits of insulating fabric hung, he selected a second shroud, pair of gloves, hood and overshoes of the transparent material, and put on the whole outfit over those he already wore. This double protection might be unnecessary; he rather thought it was. But De Soto was a cautious worker, careful of his perfectly tuned body, and he took no chances.

Ready at last for his test, he carefully connected his eighteen-inch tube to the terminals of what Crane called the devil box—a black cubic yard of insulated steel, apparently, capable of delivering a steady current at anything from one volt pressure to twenty million. If necessary, he could pass the full twenty million through his tube for a week. Having made the requisite adjustments, he released the full twenty million volts at one turn of a thumb screw.

There was no hissing crackle, no sudden splash of blinding light from the tube, no fuss or fury of any sort whatever. For all that an unskilled observer could have told, the tube was dead. Whatever was taking place in it, if anything, was far beyond the spectrum of light. If waves were being generated under the terrific impact of the electrons in the tube, ripped from the cathode by the full bolt of twenty million volts, those waves were so short that they affected no eye. Anyone except its inventor might have casually picked up the tube with his bare hand.

De Soto seemed satisfied. He disconnected the tube and turned a small screw on its side. Gradually, a thin pencil of black metal advanced into the vacuum, directly into the path that the discharge from the cathode must follow. The tube was again connected and the full current turned on and quickly off. Again there was no flash or other obvious indication that anything had happened. By a simple device it was possible to remove the pencil of metal from the tube without admitting a single atom of gas into the tube. Having removed the pencil, De Soto walked to the farther end of the laboratory, adjusted the pencil between the terminals of a huge storage battery, and turned on half the current. The metal pencil glowed from dull red to scarlet, then to pale blue, and finally to a dazzling white as the current passed through it. De Soto reached for a pocket spectroscope and studied the light emitted by the incandescent metal. Evidently the result so far was satisfactory, for he smiled. Still keeping his eye on the spectrum, he reached out with one hand and switched on the full current. There was a flash, a sharp report, and darkness. He had seen all that was necessary. The pattern of brilliantly colored lines crossing the spectrum as the incandescent metal exploded to atoms told its own story. The metal had been transmuted into a different element from what it had been before the full discharge of the twenty million volt tube struck it.

One after another De Soto inserted pencils of the metallic elements into his tube, gave them the limit of what the devil box concealed, extracted them, and subjected them to the simple, conclusive tests of flash spectrum and spark spectrum. He was already bored; the outcome was known in advance. These verifications of what he knew must take place in his hardest of all rays were but the necessary tests to assure himself that the construction of his tube was not faulty. As a mere detail in his true project, he had accomplished, on a wholesale scale, the transmutation of the elements—the dream of the alchemists and one goal of modern technology. Others had knocked the electron shells off the atoms of the elements, and still more ingenious experimenters had even tampered with the all but inaccessible inner core, the nucleus. By controlling the whole scale of radiation, from the longest radio waves to the shortest cosmic rays, in one simple, comprehensive generator, De Soto could pass up or down the whole series of chemical elements at will, as an accomplished pianist strums over his octaves. All this, however, was but the first step toward his goal. The hard, invisible, new radiations released by the transmutation of one element into another were the tools he required for his untried project.

For the moment he was lazily satisfied. After all, it was no great feat. Had he not done it when he did, some routine worker in physics must have succeeded within ten years. Literally scores were racing against one another along parallel roads to the same end. Where he surpassed what the natural development of physics would have suggested, was in his absolute control of the mechanism of disintegration. No sudden outgush of energy destroyed his apparatus in uncontrollable fury. The perfection of his technique permitted him to stop instantly any explosion of matter that might start; in fact an automatic regulator of the simplest pattern—simple, that is, to anyone with the eye for seeing nature as it is—held the incipient whirlwinds of destruction in leash.

So far all was well. To proceed farther he needed new materials. No rudimentary metal would serve his purpose. He must have highly complex compounds of a score of elements, all delicately adjusted in perfect, natural balance. The man-made products of the chemical laboratory were too artificial. If one would question nature, he must use the living things that are nature's most perfect mode of expression.

As he raised the blinds, he pondered his next step, and he smiled. Nature, or chance, he thought, had been kind to him. It had given him a perfect body and an unparalleled mind. What more could he wish? A partner with whom to share these bounties of generous mother nature. The thought of what he was about to do suddenly doubled him up in a spasm of laughter.

"Millions and millions and millions like her," he gasped, "and they don't know what is going to happen. All like her, every last one of them. How long will it last? Another ten or twenty million years? Or perhaps thirty million?" For a moment his mirth overpowered him. He was helpless, till his bitter humor died of its own exhaustion. "Thirty years," he said slowly, coldly, in answer to his own sardonic questions. "Eighty years from now every living thing will be happy. This is the end we have striven for since the days when we lived with beasts, and like beasts, in mouldy caves. Who will ever guess? Frogs and guinea pigs, Alice and the millions like her, my own children, will never know what I have done for them. What a joke!"

He removed his protecting hoods, shrouds, gloves and overshoes, and strolled over to take a last look at his tube. An exclamation of dismay burst from his lips. The crystal window of the tube glowed with a faint green fluorescence. Leaping to the devil box, he searched frantically for the faulty connection which permitted the current to leak into the box. There was no means of "killing" the whole box; it was automatic and self-contained. Until he found the leak there was nothing to be done. Not succeeding in his first dozen frenzied trials of the screw switches, he raced to clothe himself again in his protective armor. Was he too late? The hardness of the rays emitted along with that pale fluorescence was an unknown quantity until he could determine the strength of the current leaking from the box. In his confusion he had not thought of the obvious way of disconnecting the tube from the box entirely, but had assumed that some switch was not fully open. Hence his first frantic attempts now rose up to reproach him with the stigma of stupidity.

"Am I like the rest?" he gasped, hurrying back to the box. "A blundering fool after all?"

This time he found the trouble at once. The one screw switch that he could have sworn he opened was still just barely closed. For anyone working around such deadly apparatus, this trivial oversight was a blunder of the first magnitude. Approximately two million volts were still streaming into the tube.

"I should have tested that switch first," he muttered. "Am I a common fool? Well, this is a warning to do everything hereafter in the stupid, routine way these cattle use about this stable."

Once more he removed his protective garments. This time he found everything 'dead,' as it should have been. For a moment he felt old and tired. Before he realized that he was speaking, his lips had propounded a strange question.

"Who am I?" The voice seemed to speak from a forgotten world. "And what am I doing here?" Again the words seemed uneasily familiar. "Didn't I make a mistake like that before?" he continued in his normal voice. "Where was it, and when? Strange that I can't remember. Yet I could swear that I once saw a green light like that, only more intense. It brightened, and grew white. Then a black piston destroyed it in total darkness. How could that be? Only the complete destruction of radiation could give such an effect. I never did this before. What is the matter with me?"

Unable to answer, he left the laboratory, carefully locking the door behind him. Once outside in the brilliant sunshine he recovered rapidly. The cloudiness of his mind quickly cleared, and he began visualizing his immediate purpose. It was a few minutes past twelve. He stepped into the business office and used the telephone.

"Is Miss Kent speaking?" he asked when he got his number. "This is Mr. De Soto. I've been working hard all the morning, and don't fancy going to a restaurant for lunch. Can you give me something if I come out to the house? Anything you have will do."

The happy girl's reply was scarcely coherent. Yet De Soto understood the sense behind the nonsense of her words.

"Thank you, Alice,. I shall be right out."

As he got into a taxi the queer sensation again overcame

him for a moment. "Am I going soft? Bah! All I need is food and exercise."

To Alice the luncheon was nectar and ambrosia; to De Soto it tasted like lobster and icewater, which is partly what it was. The food, however, was not his chief concern. As they passed into the conservatory, he came to the point.

"Alice," he began in his resonant voice, "what happened yesterday has made me think a great deal about your future and your father's. Who would take care of you if I were to go away? No; please don't interrupt. You do need looking after, and so do I. Why can't we compromise? I have loved you since the first night I saw you. Will you marry me—now, this afternoon?"

Her answer was a foregone conclusion, and De Soto knew it. Feeling her warm young body in his arms he almost got a thrill.

They were married at three o'clock by a justice of the peace. Kent was not present, as he had disappeared into the bowels of the city in search of a job and could not be reached by telephone, radio, or prayer.

It cannot truthfully be said that love transformed Miguel De Soto, however devoutly such a consummation was to be desired. His marriage was for a definite purpose. If, toward the end, he got more than he bargained for in the way of love, it was by accident and not design.

To prepare for what he hoped to do, he took Alice on a shopping expedition as soon as they were married. She ordered whatever took her fancy in the way of personal adornment, while De Soto admitting a weakness which she had never suspected, won her bridal love completely by his own purchases. They were made in queer quarters of the city, near the market places, where live stock is offered for sale. Small animals, he declared, had been his boyhood friends, and now that he was a sedate married man he could afford to gratify the thwarted longings of years to possess a select menagerie of his own. Guinea pigs, white rats and robust frogs were his special pets, although he also betrayed a weakness for the common chicken. All this squeaking, croaking, crowing and cackling family was ordered to be delivered at once to the Foundation residence—Kent's former home—and to be installed in the appropriate pens, coops, runways, ponds and cages by nightfall. By liberal bonuses De Soto extracted a ready promise that his happy family would all be settled in the back garden by six o'clock that evening. It was a collection that would have made a geneticist's mouth water. Alice almost cried, so happy was she at the unsuspected tenderness of her husband and lover.

At six o'clock Kent returned, footsore, heavy of heart and weary after a fruitless search all day for employment, to be greeted by his new son-in-law. Alice at the moment was upstairs. With great tact and delicacy De Soto hinted that Kent had better find quarters elsewhere, for a few weeks at least. Kent was so overcome with joy to learn that Alice at last had captured the elusive, reserved young genius—and millionaire—that he fell in with the suggestion at once.

"I understand how it is, my boy," he assured De Soto, laying a fatherly hand on his shoulder. "Just let me run upstairs and tell Alice how happy I am, and I'll be off at once."

De Soto resented the "my boy" of his father-in-law, but did not show it on his smiling face.

"By the way," he said casually, "since I am now one of the family, I shall pay all bills." He unobtrusively slipped a handsome check into Kent's hand. "As you belong with us by rights, it is only fair that I should take care of your hotel expenses." De Soto was not mean. Money meant nothing to him, and he cared a little less than nothing for the things that can be bought by money or genius. Crane was right when he told the President that De Soto was too brainy to take pride in anything.

While Kent was upstairs bidding Alice adieu and almost

crying with her over this happy issue out of all their afflictions, De Soto paced the dining room carpet like a trapped tiger. For the first time that he could remember he felt maddeningly, stupidly ill. A hot prickling tingled over every inch of his skin like needle points of fire. It had first come on, faintly, while he was buying guinea pigs with Alice. Although he gave her no hint of his distress, it had required all of his self-control to act as if he were in perfect health. To one who recalled only the easy sense of wellbeing of a young and healthy animal, the first experience of illness was mental torture not to be endured. De Soto was out of his depth. What should he do? Consult a doctor? He would ask Kent the name of a reliable man the moment he came downstairs.

Taking a grip on himself as he heard Kent's step, De Soto stopped his feline pacing and stood rigidly still.

"Who is your family physician?" he asked in a level voice.

"Brown," Kent replied, somewhat surprised. "Most of your staff consult him. Not feeling unwell, I hope?"

"Oh, no. But I thought Alice may have been doctoring with the wrong man. She has looked rather pale the last few weeks."

"That will all mend itself now," Kent assured him. "You are the right doctor for her."

The moment he was gone, De Soto bounded upstairs and tapped on his wife's door. "It is Miguel."

"Come in," she cried in a low voice.

"Alice," he began, "can you ever forgive me for running away and leaving you to dine alone? I have just remembered that I left a switch closed in my laboratory. I must go back at once, or the whole place may be wrecked."

"Can't you telephone to someone?"

"No. It is too dangerous."

Her face went white. "Let me come too," she begged.

"That would be worse than ever. I know exactly what to do. There is not the slightest danger to me. Anyone else—"

There was no need to finish the sentence; he had produced the intended effect. "Don't expect me till you see me—I may be hours."

She kissed him passionately.

"I won't be a drag on you," she declared, "even on our wedding day."

For reasons of his own, he went out by the back door. On his way through the service port, he hastily emptied half a sack of potatoes into a box and tucked the sack under his arm. In the patio he found that the assorted pets he had purchased were all comfortably housed in their respective quarters.

"I may as well do two things at once," he muttered, slipping two large frogs and a pair of guinea pigs into the sack. "She might ask questions if she saw me taking these away in the morning."

Unobserved by any of the servants he got the sack into Alice's car. His immediate destination was Doctor Brown's house. At a drug store he learned the doctor's address.

The doctor was just sitting down to a bachelor's dinner when the housekeeper announced that Mr. De Soto wished to see him at once. Brown rose with alacrity. He and De Soto had not met. The doctor, however, felt that he knew De Soto better perhaps than the young man knew himself. He found his caller in the study.

De Soto went straight to the point.

"Mr. Kent recommended you to me. For the past three or four hours my whole skin has felt as if it were on fire. Will you examine me? I may tell you that Miss Kent and I were married this afternoon."

"No wonder you are over-anxious about yourself," the doctor laughed, concealing his shock at the news of Alice's marriage. If De Soto was the man Crane thought him, poor Alice had been undeservedly punished. Brown had always liked her. Single or married, he silently resolved, he would stick by her.

"Is my condition likely to be serious?" De Soto asked, his vibrant voice growing husky with repressed animal fear.

"Probably not. I treated a similar case successfully a short time ago."

A curious change came over De Soto's eyes. For a moment they might have been those of a wild beast trapped and about to be killed. Brown caught the flash. It convinced him that Crane was not far wrong in his estimate of this young man.

"What was the patient's name?" De Soto demanded. His question was harsh and hoarse with fear.

To Brown it was evident that De Soto suspected Crane of having used the forbidden tube secretly. The doctor rose to the occasion.

"Oh," he replied, "the case I speak of was six months ago. What was the man's name? Let me think. It was before you came here. He was an assistant at the Foundation and got into trouble. You may have heard of him committing suicide. I've got it. Bork. That was the man."

"Aren't you mistaken?" De Soto asked in a voice which he did not recognize as his own. The question came from his lips involuntarily, as if some personality deeper than his own were expressing a doubt.

"I think not. Why do you ask?"

"Ask what?" De Soto rubbed the back of his hand across his eyes.

"You've been working pretty hard of late, haven't you? Take my advice and lay off for a spell. You just had a slight lapse of memory then. Well, the first thing is to ease your skin. We can do that here, in my bathroom."

Brown himself gave De Soto his bath. The moment the patient was out of the tub, the doctor hustled him into the dressing room and rubbed him down thoroughly with disinfectants. He was not going to lose the priceless bathwater this time, or have any third party see it pass from pink to crimson.

"You will be comfortable for an hour or two anyway. I suppose you won't want to go home until you feel sure you're cured. Did you tell your wife anything?"

De Soto confided the fiction of the danger of the laboratory, but saw no necessity for mentioning the sack and what it contained.

"Fine. Go to a hotel and spend the night there. Repeat the treatment every three hours. I'll give you a prescription for some stronger stuff. Telephone to your wife that the job at the laboratory will keep you there till ten or eleven tomorrow morning. Come and see me again at eight."

Promising to carry out the doctor's instructions to the letter, De Soto left. The desk clerk at the hotel addressed him by name, feeling highly honored to have as a guest the young inventor whose picture was always appearing in the papers. De Soto got the best room and bath in the hotel. Morbidly concerned about his health, he did not wait three hours to repeat Brown's treatment, but did it at once with drastic thoroughness. Then, cold with fear, he lay down to torture himself for two hours with unreasonable fancies. Brown had assured him the other man recovered quickly and easily. Would he? His fear of bodily discomfort was not cowardice, but simply the natural reaction of an animal experiencing its first pain.

The two hours passed without the slightest recurrence of the symptoms. Encouraged, De Soto repeated the bath and disinfection, and lay down again, this time with dawning hope. Luck stayed with him. And so it went, with complete success, till two in the morning. Feeling that he was free of his trouble for good, De Soto dressed and left the hotel. The numerous bathings and rubbings had made him feel like his old self, full of energy and eager for work. He got his car and drove to the laboratory. The sack had not been molested during his stay in the hotel.

8

TRAPPED

SHORTLY BEFORE MIDNIGHT, WHILE DE SOTO in his hotel room was busy with his prophylaxis, Crane descended upon the doctor to prepare for the proposed raid on the forbidden laboratory. Brown knew that he was taking his reputation, if not his life, in his hands by sharing Crane's somewhat foolhardy enterprise. Nevertheless he was determined to go through with it for scientific reasons as well as for the sake of friendship.

Crane found the doctor peering through his binocular microscope.

"Anything new?" he asked.

"Not exactly. The same protozoa that you contributed to the cause of science. This really is most extraordinary. Old Wilkes would give his right eye for one look at these. They mustn't be exposed to the light too long, or they'll vanish into nothing."

In answer to Crane's rapid fire of questions, the doctor explained how he had secured his fresh supply of protozoa. The announcement that De Soto and Alice were married was received in cold silence. What could be said? The time for talk was past.

"De Soto has blundered," Crane hazarded finally. "What it took thirty hours for the two million volt tube to develop on my skin has shown up on his in half a day. This must have happened after I left this morning. My hunch is that he doesn't know what he is doing. Well, shall we be going?"

"If you insist, we may as well." The doctor pocketed his flashlight and a small medicine case. "I'll go and give Bertha her sleeping potion and join you in front."

Forty minutes later the two rash men were outside the door of De Soto's laboratory. Brown carried a large paper market bag in which the drugged brown hen reposed limply and silently. There was some slight difficulty at first in forcing the new key into the lock. Crane began to swear softly.

"Hadn't we better give it up?" Brown suggested. "A superstitious man would say our trouble in forcing an entrance is a sign from Heaven to quit."

For a moment Crane was inclined to agree. His square-jawed obstinacy, however, persisted.

"There," he whispered at last, as the key turned in the lock. "In with you!"

Before turning on the lights, Crane cautiously felt his way from one window to the next, making sure that the iron shutters had been closed as usual for the night.

"All safe," he announced, rejoining Brown by the door, and turning on the lights. "Now to find out what friend De Soto thinks he is doing."

"Your key?" Brown suggested. "Hadn't you better leave it in the lock?"

"No. The watchman is not due this way for nearly two hours yet. But suppose he were to come round out of his regular beat. If he found the door locked from the inside he would ring till he was let in. Otherwise he would just open the door, turn on the lights, look around from here and lock up again. You must stand here and switch off the lights if you hear anyone coming. I shall duck into that closet—where the insulating togs are hung—and wait till he goes away again. After turning off the lights, you sneak round behind that steel cabinet and stand as close as you can to the window. The watchman won't see you from where he stands. Take the hen with you, of course."

To forestall the unexpected, as all trained scientists do, the conspirators rehearsed their parts six times before attempting any experiment. While Brown switched the lights on and off, Crane practised his disappearing act between the devil box and the closet. The doctor for his part managed to turn off the lights and vanish behind the steel cabinet almost in the same moment.

"Safety first," Crane grinned when the rehearsal ended. "The unforeseen always happens. Turn off the lights till I give the word."

As a last precaution, he unbolted the iron shutters of the window by which Brown was to stand in case of danger, unlocked and raised the window, and finally closed the shutters without rebolting them.

"If we're caught, you fling open the shutters and step out of the window. Then beat it."

"And leave you to face the music? What do you take me for?"

"A man of common sense. Don't argue. It is as much to my advantage as it is to yours not to be caught four-handed, as it were. The President will take care of me. Your reputation would be gone forever. Now, do as I say. I'm the captain here; you're a buck private in the rear rank."

After much further argument Brown consented. The point that finally won him over was quite unanswerable. Crane refused to start the experiment unless the doctor first gave his word as a gentleman to obey orders to the letter.

"We don't know what sort of rays De Soto's tube generates," Crane remarked, reaching into the closet for a protective suit. "Let us take no chances. Put on double armor."

Their preparations at last complete, they hopefully set about the experiment for which they had come. Bertha, still soundly drugged, was left in her sack by Brown's emergency window. If the tube generated nothing more penetrating than even hard X-ray, the unsuspecting hen would be amply dosed where she lay. But Crane, examining the curiously compact mechanism of De Soto's little masterpiece, had an uneasy feeling that the tube could emit radiations infinitely more dangerous than the most penetrating rays known to human science. Trusting that their double protection was sufficient, they tried to connect the stocky little tube to the black devil box.

For five minutes Crane fumed and fussed at the ridiculously simple terminals.

"Better take up your station by the lights," he snapped irritably to the helpless doctor. "I can't seem to make it work. We may be fooling here till daybreak."

Brown humbly retired to the door and, to reassure his exasperated collaborator, lightly laid his fingers on the buttons controlling the lights. Crane had not the least idea of what he was doing. Accustomed to the usual sputtering of ordinary tubes he naturally imagined that nothing was taking place in the silent, dead-looking apparatus before him. The transparent gloves, thinner than silk, seemed to interfere with his manipulations. With a gesture of irritation, he started pulling them off, when Brown sharply stopped him.

"Don't do that! How do you know that box is safe?"

"The box can't do any harm. It's the tube that counts. Why doesn't it glow?"

"Don't take off your gloves! It may be—."

The doctor's expostulation was cut short by a drowsy voice from the window that seemed to ask "What?"

"The lights!" Crane muttered tensely.

Instantly the laboratory was plunged in total darkness.

Brown recovered his nerve first.

"That was only Bertha coming to," he laughed. "Shall I switch on the lights?"

Crane assented, and once more began tinkering desperately

with the connections. Barely had he started when the lights went off again. For the moment he forgot his own instructions to Brown.

"What's up now?" he fretted.

"Steps! Duck!"

As he shut the closet door noiselessly after him, Crane heard a key being inserted into the lock of the laboratory door. Brown was already in his station by the window, praying that Bertha would not continue talking in her sleep. The door opened, and the lights were turned on. It was De Soto carrying his sack. Neither Crane from his coal-black closet, nor Brown from his station by the window could see who the intruder was. The doctor wondered why the supposed watchman did not turn off the lights and go away. To their horror both men heard the door being closed, the key turned in the lock, and the sound of confident footfalls advancing into the laboratory. Brown had the additional discomfort of knowing that the lights were still on.

What followed was like a hideous nightmare to the three participants. In four minutes history was made on a scale that would have paralyzed the minds of at least two of the protagonists had they but dreamed what their foolhardy tampering with forces beyond their childish understanding would precipitate. Neither Crane nor the doctor saw in De Soto's outburst of fury anything more significant than the ungovernable rage of an overwrought man magnifying a real, but not very important wrong, into a cosmic disaster. Their crass bungling had unchained the devil. If any justification for De Soto's career be possible, it resides in the history of those four epoch-making minutes. According to his own account, he intended something quite different for the world, and for Alice in particular, from his actual campaign. We have only his word for all of this. But, in the absence of conclusive evidence to the contrary, it is simplest to assume that De Soto was not a liar.

The historic episode began with a hoarse, despairing cry

from De Soto. In one amazed glance, as he walked toward the black devil box, he had noticed that the tube was fully connected as efficiently as if he himself had linked up the twenty million volts to the evil fiend of his own devising.

"I'm a fool like the rest," he wailed, dropping the sack with the frogs and the guinea pigs. "I left it on!"

He darted for the closet to fetch himself a suit of the transparent armor. Crane heard him coming, and squeezed himself into the farthest corner behind three of the dangling shrouds. De Soto groped for a shroud, gloves and overshoes without looking at what his hands grasped. Shouting incoherent nothings he got himself into a single suit, and darted for the devil box to disconnect the tube.

"I should have done this first," he raved, realizing his blunder too late. "Fool, fool, fool! What am I?"

Bertha brought the tragedy to its climax. As De Soto's lightning fingers disconnected the tube, a final surge of energy ripped the innermost cells of her body apart. Although she was only a brown hen, that exquisite pain gave her for a fraction of a second a voice that was three parts human. Her croaking shriek rose shrilly above the unnatural cries of the outraged guinea pigs and frogs in the sack, whose innermost sanctity of life had also been violated in that abrupt surge, and froze the fingers of the man blundering at the tube.

"What was that?" he yelled, his voice the cry of a lost animal facing death.

As if in answer to his question an iron shutter seemed to open of itself; a black mass hurled itself into the blacker night, and the wailing shriek of the outraged hen receded into silence, and died. Brown had escaped with his booty.

De Soto found himself staring in a dream at the fatal work of his too penetrating intelligence. That it had ruined him was sufficient for the moment. That his own imperfect mind, as he thought, had delivered him up to failure in its worst form, was an ironic jest that cut deeper than mere failure. His memory began to reassert itself. Surely he had disconnected the tube before quitting the laboratory? A clear visual image of the tube as he had left it, flashed on his retina. He rushed to the open window and stared into the darkness. Who had robbed and betrayed him? Crane? De Soto began shouting hoarsely for the watchman who, having seen the light streaming from the window, was already running down the walk to the laboratory.

The watchman as a matter of course began a systematic search. Within a minute he had found Crane.

"All right," the latter remarked dryly, stepping into the light and confronting De Soto. "I guess you've got me. What are you going to do about it?"

His story was already made up, such as it was. In preparation for it he had stripped himself of his protecting gloves, shrouds, hoods and overshoes, and hung them up in the closet before the watchman opened the door. De Soto regarded the suspect somberly before replying.

"How did you get in?" he demanded.

"I was passing here—having a look at things from the outside for old times' sake," Crane grinned, "when I noticed the open window and came in to investigate."

"You came in alone?"

"Presumably. The man who vanished in such a hurry when you unlocked the door must have left the window open. I judge he must be pretty well acquainted with the general layout of this laboratory. Otherwise he couldn't have found his way about in the dark."

De Soto affected to credit this theory.

"You had better tell the proper authorities in the morning," he said, watching Crane's face narrowly.

"Of course, if you think it necessary. I will tell the President of the Board, if you like, as soon as I can get hold of him."

"And you will agree to abide by his decision as to what

is to be done?" De Soto suggested with a malicious smile. Crane nodded. "Then," De Soto continued, "I shall have to ask the watchman to search you. A mere formality," he smiled, "so that I may be able to assure the President that you were only safeguarding the interests of the Foundation like a loyal alumnus."

"Rather rough on me, isn't it? What if I object to being searched without a proper warrant?"

"None is necessary. You were trespassing. Search him."

The key was found at once. Although pretty far gone, the game was not yet lost. Crane pretended to take the damning discovery as a matter of course.

"Yours?" De Soto asked.

"Of course."

"Then you had a duplicate?" Receiving no reply, De Soto explained. "To make sure that you would attempt nothing rash after your dismissal yesterday morning, I asked at the office before going home whether you had turned in your keys. They told me you had. This insanity on your part confirms my suspicions. You were discharged, as you doubtless guessed, because I had a strong feeling that you were spying on me. Now, if you will tell me who your confederate is, I shall take no action against you. Who is he?"

"I don't know what you are talking about. That duplicate key must have been in my pockets for weeks."

"It is too new," De Soto pointed out coldly, holding it up for the watchman's inspection. "For the last time, will you tell me who was in here with you?" Crane's obstinate silence seemed to infuriate his inquisitor. "You refuse? Then I shall turn you over to the police."

"What good will that do you? The trustees will believe my story—to avoid a scandal, if for no other reason."

"You think so? Possibly you know them better than I do. Let me think a moment." As if trying to make up his mind, De Soto began pacing back and forth in front of his devil box.

At last he appeared to reach a decision. "Close that window, fasten the shutters, and go about your rounds," he directed the watchman. "Lock the door after you. I will be responsible for this man till you look in again."

"Now," he began when they were left alone, "there are no witnesses. We can speak the truth without fear of the consequences. How long had the tube been connected to the box when I came in?"

This tempting invitation to give himself away completely did not appeal to Crane.

"How should I know? Whoever was in here may have been tinkering with your apparatus for three minutes or three hours."

"So you refuse to talk? Very well; I shan't press you. Amuse yourself till the watchman comes round again. I must see what damage has been done."

Turning his back on his prey, De Soto strode toward the evil black box. For half a minute Crane did not foresee his intention. Only when De Soto began rapidly making the connections necessary to operate the tube did the truth flash upon him. He was absolutely without protection against whatever fiend De Soto, himself sheathed from head to foot against the rays, might release. The memory of the unnatural cry which Bertha had emitted when Brown—also protected as Crane was at the critical moment—snatched her with him in his flight, roused every instinct of self-preservation in the doomed man. One terminal was already connected. De Soto's nervous fingers were about to close the circuit by connecting the second, when Crane hurled himself upon his inhuman enemy.

The unexpected impact catapulted De Soto against the black box and flung him violently to the concrete floor. Before Crane could fall on him, he had rebounded like an enraged tarantula and leapt to the farther side of the box. Vaulting the box, Crane tried to seize the desperately cool devil sneering into his face. De

Soto kept him off easily with one hand, while with the other he felt for the second wire dangling above its binding screw. Crane's wiry strength was no match for the perfect machine of bone, muscle and brain opposing him. The free hand made the connection and began groping for the small button switch that would release twenty million volts to surge into the tube. The operator knew what the consequences to the other must be; the intended victim could not even guess, except that they would be evil. The all but human cries of the hen, and frogs and the guinea pigs seemed to echo again through the laboratory. What would his own cry be like?

Instinct saved him. Powerless to prevent the groping fingers from finding their mark, Crane ripped the hood from De Soto's head in one convulsive movement with all his strength.

"You know, then!" De Soto shouted, making a leap to recover the hood.

"I know you're crazy," Crane jeered, eluding him. He saw his chance and took it. Before De Soto could pounce upon him, he had seized the tube and hurled it to the floor. If not utterly ruined it was out of commission for at least the remainder of that night. Panting from rage, De Soto stood staring in speechless hate at Crane. At last he got his breath.

"You fool," he gasped. "The next ten years in the penitentiary."

"Not if I know it," Crane retorted. "Any jury would let me off if I told them what you were trying to do to me."

"What was I trying to do?" De Soto demanded with deadly calm.

"Nothing for the good of my health. That's all I know, and it is enough."

"You're insane. I try to find out what damage has been done to my apparatus and you attack me like a madman. Explain that to your jury. Also tell them that you deliberately wrecked my tube. Remember that you were discharged on the suspicion of having tampered with it already. Will you wait for the watchman, or will you come with me to the police station and surrender yourself?"

"Why not compromise? Suppose we talk these things over with the trustees in the morning. After all, you are only the Director, you know. This laboratory isn't your private property. That tube I have just smashed and everything else in here belongs to the Foundation. I'll meet you in the president's office at nine o'clock."

"So the trustees hired you to spy on me?"

"I'd be likely to tell you if they had, wouldn't I? Look at the common sense of our row for a change, and give your imagination a rest. You lost your temper and went clean crazy. Then you tried to give me a dose of something you don't like yourself. It probably wouldn't have killed me. You're not so crazy as that. But it might have done something worse. You know best. I'm willing to go before the trustees or the police, because I shall suggest that they find out exactly what you are trying to do with your short waves."

To Crane's astonishment, De Soto began to rock with laughter.

"I will tell you," he confessed. "The trustees are good business men, but they need educating. I planned to educate them. Now I have changed my mind. They were too fond of money, I thought, and they used my brains to flood the world with trash that only fools would want and only imbeciles pay for. Without me they would still be poor. Now they dream of owning the world. And but for my silly inventions the public would never have dreamed that it could want the stuff it buys. They asked me for rubbish because they could imagine nothing better to want, and I gave it to them with both hands as I would shower idiotic toys on a half-witted child. Like a fool myself I thought it would be a great thing to show all of them the one thing that every rational animal should crave."

"Which is?" Crane interrupted.

"Why should I tell you? Your prying incompetence may have wrecked my work. And you, like all of the bunglers earning their livings here, pass for a man of more than high average intelligence. Could you be educated to want what I planned to give the trustees and all their dupes? No. Nor can any living man or woman. So I shall change my plan and glut you all with what you crave. You deserve nothing better. Tell the president whatever you like. You can go."

"Before I do," Crane replied grimly, "let me tell you something for yourself. I don't understand what you are trying to do, and your high theories pass clean over my head. But I am sure of one thing. You are lying. By dropping your charges against me, you hope to pull the wool over my eyes. Well, you won't. I shall tell the trustees nothing."

De Soto laughed indifferently.

"Here's your key," he said, restoring Crane's duplicate. "Let yourself out. But don't try to come back, ever again. The next time I may have changed my mind."

When Crane was gone, De Soto picked up the tube and examined it critically. The damage could be repaired in two or three days or, if necessary, a new tube might be constructed in five weeks.

"I shall need batteries of these all over the civilized world," he mused. "Then the golden age will dawn."

He locked up and walked slowly through the cheerless mists of the early morning, thinking gloomily of his bride. She would have waited up all night for him, he guessed, in spite of her assurance that she would not be a drag on him in his work. All of his grandiose projects for the human race had gone glimmering through no direct fault of his own. Any ordinary man would have said "a good job too"; for no such man could have foreseen as clearly as did De Soto the inevitable end of the race from its present state.

Passing a dingy restaurant, he suddenly realized that he was faint from hunger. Not until he had taken a seat at a slovenly table did the full depth of the profound change which had overtaken him in the past three hours register on his consciousness. An untidy waiter in a soiled white apron came to take his order.

"A ham sandwich and a cup of black coffee," he said, without looking up.

"I don't need to ask Crane or anyone else how long that tube was connected," he thought bitterly as he sat sipping his coffee. "Six hours ago this stuff would have stuck in my throat. Now I need it. I'm not well."

Indeed he was not. Idly picking up the greasy menu, he began listlessly reading through the list an item at a time. It did not even occur to him at the moment that this was not his "natural" way of getting the sense out of print. Exasperated by this unaccustomed difficulty in following the meaning of what he read, he finished his coffee, flung down a dollar, and left the place. The cool air refreshed him.

"How stupid," he muttered. "I forgot to bring the frogs and the guinea pigs." His "natural" mentality began to reassert itself. He hurried back to the laboratory and got his sack with the four animals. Outwardly they seemed normal. What were they like inside? Only time would show. The prospect of an interesting experiment cheered him up, and he went straight home, to find Alice anxiously waiting for him in the breakfast room. Again a slight lapse of memory warned him that he was unwell.

"What have you in the sack?" she asked, when their greetings were finally concluded.

"Oh," he lied readily, "some new pets. I saw them in the window of a Mexican restaurant and bought them. Two guinea pigs and a pair of frogs."

Alice was enchanted and De Soto, with a curious twinge as of some forgotten instinct stirring within him, noticed that she was charming. When had he been charmed by a girl before? He could not remember. His impersonal mind was, however, still uppermost. Before sitting down to a light breakfast with his bride, he carefully housed his "new" pets, each pair in a separate pen away from all others of the same kind.

"Alice," he began after breakfast, "I have a queer sort of honeymoon to propose. At first I had hoped that we might get away for a week or two, but an awkward turn in my work has put a trip out of the question. Suppose you come down to the laboratory when you have nothing better to do and watch me work? You can bring a book to pass the time when I'm too busy to talk."

She was more pleased than if he had suggested a six months' pleasure trip to the most frivolous playgrounds of Europe. As her husband looked tired, however, after his strenuous night in the laboratory—as she supposed—Alice insisted that he spend at least the morning in bed before they began this most delightful of all honeymoons.

When Crane left the laboratory a free man owing to De Soto's generosity, he went to the nearest telephone booth, learned that Brown had arrived home safely with Bertha, and made a dinner engagement at the doctor's house to talk things over the following evening. Both men were too fagged to think clearly until they had enjoyed a long sleep. Brown made arrangements with a friend to handle his practice and went to bed, determined to sleep ten hours at least. Before turning in, however, he took a pint sample of the bath water which De Soto had left in the tub, wrapped the bottle in several thicknesses of black paper, and left the package with a note for his housekeeper, requesting her to send it, with his card, to Professor Wilkes by special messenger the first thing in the morning. On his card he wrote: "Professor Wilkes. Please examine this sample microscopically at once. The trick to keep it from decomposing is to exclude all light. I will be at home after seven tonight." This time he felt sure of himself.

The professor, duly instructed by Hayashi's microphotographs of the protozoan fish parasites, would not pour the interesting sample down the sink. Brown's forecast proved right. At four o'clock that afternoon the housekeeper had almost to use force to turn Wilkes away. The doctor, she asserted, had given the strictest orders that he was not to be awakened till five o'clock. The professor left, lugging his heavy brief case, which the housekeeper erroneously mistook for a salesman's portmanteau. At six he was back, this time not to be denied admittance. While Brown was shaving and dressing, and apparently taking his own time about both, Wilkes dumped the contents of his bulging brief case on the hastily cleared study table and displayed his astounding evidence—to him it was no less—like an elaborate variety of solitaries played with a dozen packs of cards.

At last Brown entered.

The professor, he decided, had suffered in silence long enough.

"Look at that!" Wilkes exclaimed with a dramatic gesture toward his massive game.

"I don't have to," the doctor retorted. "I saw all that through the microscope before you threw away my first sample."

"But they form a perfect series," Wilkes expostulated, "from the lower species to the highest possible, and only the first half dozen of them recorded. Over a hundred and eighty types of protozoa new to science at one swoop! Where did you get them?"

"Where did Hayashi get his?"

"Diseased fish. But that all cleared up months ago. There has never been anything like this in the history of biology. Where did you find these?"

"I made them," Brown replied coolly, not expecting to be believed.

He wasn't.

It developed that Wilkes had spent the day making crude sketches, as fast as his fingers would work, of the curious life—or rather death—in the pint sample which Brown had sent him that morning. Most of the sketches were mere rough outlines. Some, however, exhibited considerable detail. These marked every fifteenth or twentieth place in the long series into which Wilkes had arranged Hayashi's photographs, a few of Brown's sketches, and his own. The effect, as the eye ran rapidly down the entire series, was roughly like that of a motion picture of a rosebud opening out in full bloom. Development of some sort, not mere growth, was evident. The sizes of the creatures depicted remained approximately constant; their complexity, however, increased with beautiful regularity to its climax, reaching a maximum at about two-thirds the total distance from the beginning of the series, and falling steeply down the decline to degenerated simplicity of structure at the end. It was as if a whole race of living things were maturing to its peak before their very eyes, and toppling to its inevitable extinction even as they watched.

"Well," said Brown. "Do you believe now?"

"I will, when you tell me what to believe. I can't doubt my own eyes."

"Nor your own common sense?"

"What has common sense to do with it? We are face to face with a new fact of nature."

"That's what I thought the first night I saw all this happening in a drop of mist. But there is an explanation. It is so simple as to be almost shocking. Haven't you guessed it?"

"More or less hazily. The time scale is all wrong. Impossible, I should say, if—"

"If you hadn't seen it yourself. Excuse me a minute; that must be my friend for dinner. You'll join us, won't you? It's Crane, the X-ray man."

"Does he know anything about all this?" the cautious Wilkes demanded, making a move to secrete his drawings.

"He should, as it was off his skin that I collected my first specimens."

"Introduce me at once!"

On being presented to the excited professor, Crane modestly denied any design in his startling contribution to biology.

"Doctor Brown," he concluded, "will probably have something more exciting to show you soon. By the way, doctor, did you give Bertha a bath after you got her home?"

"Great Scott! I clean forgot the possibility of her being infected as you were. Excuse me a moment. Dinner won't be ready for twenty minutes yet anyway." He dashed out to bathe the unsuspecting Bertha.

Left alone with the professor, Crane submitted resignedly to a barrage of questions. How had he ever suspected the existence of these teeming protozoa on his skin? Easily enough, Crane explained, adding that the professor himself would have been in no doubt under the circumstances. Venturing no theory, he went on to state briefly the beginning of the whole story—his thirty hours' exposure, spread over several weeks, to the hard X-rays generated by his two million volt tube, the suddenness with which the intolerable itching began, and the immediate relief when the superficial cause was removed.

"You are positive that your tube generated nothing but hard X-rays? Well," the professor admitted on receiving Crane's fairly confident assurance, "it must have been the prolonged exposure that started the explosion—on your skin, I mean. None of the other biological workers with mere X-rays ever produced such results. Not that they tried, however; although now I fail to see why it never occurred to some of them to do just what you did accidentally. Of course, there would be a delayed, cumulative action under proper doses of the rays spread over a long interval. The sum of all the doses applied in one shot might well be fatal; it certainly would have a different effect from repeated applications of small amounts. Isn't that so, Brown?"

he appealed to the doctor who had just reentered the study.

"Probably not," Brown laughed. "But I confess I did not hear your argument."

Over the dinner table, Wilkes elaborated his not unreasonable theory, letting his soup cool until the diplomatic maid removed it untasted.

Brown did not disagree. In fact he pointed out a similarity between Wilkes' theory and the standard treatment by X-rays, whereby a strong beam that by itself would seriously injure healthy tissuc, is split up into ten or more parts all focused on the desired inaccessible spot. However, he was less interested in Wilkes' guess as to what he termed the "explosion" of the protozoa than in what the nature of that explosion itself might be.

The professor was game.

"Don't laugh at me," he began, "and for Heaven's sake never tell any of my colleagues that I ever talked such fantastic nonsense. Well, here goes. It's an old story now how Muller, Dieffenbach and others first managed to produce permanent modifications in certain living flies, that were transmitted for generation after generation to the remote descendants of the original flies. You recall how it was done; the perfectly normal flies were exposed to X-rays, and then carefully segregated and watched while nature took its usual course with flies. They increased and multiplied. But some of the sons and daughters had curious defects of the eyes and other peculiarities from which their parents did not suffer. The sons and daughters were encouraged to mate without having been treated by the rays, as their parents had been. Their offspring inherited all the acquired characteristics. Thus it went for generation after generation; the artificial modifications initially produced by the rays were passed on from father and mother fly to son and daughter fly, precisely as if the first freaks were the natural offspring of their parents—which they were not. It was as remarkable, in its own way, as if a war veteran with only one arm should have a son with only one arm

—and the same arm, right or left, and the son in his turn should have a son or daughter with the same defect, and so on for hundreds of generations."

"Don't you want any dinner tonight?" Brown interrupted, as the maid was about to make off with the professor's unviolated chop.

"There's only a salad and cheese with black coffee and crackers after this."

"Dinner? What's dinner in a crisis like this? Evolution has gone mad before my very eyes. Here," he called after the maid, "please bring back my plate. One must eat, even in a lunatic asylum. Now," he continued, firmly spearing his chop, "consider what all this means. Take the human race, for instance. We're mammals; you admit that. And what are mammals, ultimately, but an offshot of the reptiles? How did they shoot off in the first place?"

"Don't ask me," Crane muttered guiltily, as the professor fixed him with a flashing eye. "Brown ought to know."

"He ought to. But does he? No? Well neither do I," Wilkes exclaimed, evidently well satisfied with himself. "But I have a theory—no, not now. Later, when we get to the bottom of your new protozoa. What do the biologists tell us?"

"You ought to know," Brown suggested. "Don't you make your living at biology?"

"I do know!"

"The mammals sprang from the reptiles by a mutation—a sudden change of species."

"Rot," Crane commented tersely and incisively, with the superior wisdom of the physicist accustomed to manufacturing theories in the evening to be thrown overboard in the morning.

To his great disgust, Wilkes unexpectedly agreed with him.

"Of course it is rot," Wilkes assented. "I know even better than you that mutations explain nothing; they merely give a fancy name to the fact we are trying to understand. Evolution by jumps, instead of slow, continuous growth—there's another statement of the same thing. What I want to know," he exclaimed, bringing his fist down on the table, "is what causes these jumps. The physical reason—not a restatement of the problem. Something suddenly took place in the germ cells of the reptiles, and they brought forth strange creatures—no stranger than those artificial flies with the queer eyes—that later evolved into your ancestors and mine. Tell me that, and I'll rule the world!"

"Shall we tell him?" Crane asked with a dry smile.

"I haven't the heart," Brown replied in the same vein. "He might pour us all down his kitchen sink."

The debate lasted well into the night, and, like most battles with words, settled nothing. The real debate had not yet begun. Things, tangible and real, were presently to play their part in the argument. Nevertheless, the trio did succeed in forming a not unreasonable guess as to what the professor's interesting series of drawings signified. At two o'clock in the morning, when the little party broke up, they agreed to stick to the problem until they could control the protozoa at will. The professor's parting cry of triumph was to the effect that he now held the key to evolution in his hand.

"Better throw it away, then," Crane remarked. "I got caught this morning with a key that I had no business having, and I'm afraid I'm in for a peck of trouble."

On the way out, after Wilkes had gone, he briefly told the doctor of De Soto's truce.

"I don't like it," Brown remarked. "Especially his attempt to force you to tell him how long his tube had been running when he came in. De Soto has been fooling with something he only half understands. I guess that we caught him in his own trap."

"If so, we had better shoot him before he breaks out again. He's a bad egg. Good night; see you tomorrow."

9

BERTHA'S BROOD

SOME THREE WEEKS AFTER BROWN'S DINNER party, a puzzled electrical engineer in New York sat reading and re-reading the most extraordinary letter that any human being ever received. The engineer was the once celebrated Andrew Williams whose early patents on high-power transmission remade the wholesale electrical industry and founded the colossal fortune of the now defunct Power Transmission Corporation—P.T.C. as it was known in its prime.

Vice-President Williams' brain had made P.T.C. both possible and prosperous; Miguel De Soto's better brain had made it both impossible and bankrupt. The decline of P.T.C. began when the Erickson crowd captured all long distance, high tension power projects with their new principle of electrical insulation. From decline to ruin was little more than one stride, and P.T.C. took it. Overnight, when the first great advertising campaign of the Erickson began to bear plums for its sponsors, and thistles for its competitors, the stock of P.T.C. fell from 180 dollars a share to 14 dollars and 50 cents. It was a washout, and the unfortunate corporation was drowned. All that remained was to wind up the affairs of the corporation—with the help of the somewhat unsympathetic courts—and start all over again. The first would take from six months to a year. What the reorganized P.T.C. should manufacture was a mystery. Williams favored television sets, but the rapidity with which the Erickson Foundation relegated successive improvements in that field to the National Museum of Arts and Sciences, made the Vice-President's associates pessimistic and chary. They for the most part advised a complete clean-up of the business and a general retirement on the wreckage for all its officers.

The letter which caused Williams such bewildered astonishment was thirty pages long, typed in single space, and anonymous. No water-mark or other identification betrayed where the paper might have been purchased. The paper itself was rather peculiar for a business letter. It was thin, light brown wrapping paper, such as is commonly used in department stores for doing up parcels, cut to the standard typewriter size. The cutting apparently had been done with a sharp penknife. Although the typescript was plainly legible, the marks of numerous erasures on every page indicated that whoever had operated the machine was no skilled typist. The general appearance suggested that the entire thirty pages had been painfully pecked out a letter at a time. As a last, significant detail, the type had all the earmarks of that from a practically new typewriter.

Most sensible persons consign anonymous letters to the fire, if one is handy; if not, they tear the letter into small pieces and entrust it to the wastebasket. Williams, on looking for the signature and finding none, was tempted to be sensible. The opening sentence of the letter arrested his attention, however, and he read it breathlessly to the end.

"Sir," the letter began, "I herewith present you with the infallible means for recovering all of your recent losses and regaining your monopoly over the power transmission industry at no cost to yourself."

The letter concluded with the suggestion that Vice-President

Williams at once patent everything of value in the detailed specifications.

"My purposes," the anonymous writer asserted in a postscript, "are purely humanitarian and educational."

For the twentieth time Williams scrutinized the large Manila envelope in which the letter had come—unfolded. Only his own name, with the words "Personal and Important" added, all in the same kind of typing as that of the letter, offered any clue to the sender. Obviously no detective could hope to trace the letter from these data alone. Williams rang for his secretary.

"When was this envelope delivered?"

"I couldn't say. The office boy laid it on my desk at eight o'clock this morning with the rest of the mail from our own box."

The office boy remembered taking the large envelope from the mailbox with the rest. There the clues ended. The Vice President again summoned his secretary.

"If anyone asks for me, tell him I have gone to Washington, D. C. I'll leave my hotel address at the information desk in the U. S. Patent Office in case of an emergency. Don't expect me back for a week."

Williams had been a great inventor in his younger days, and he knew that noble game from alpha to omega. Unless some crank were hoaxing him by passing off as a free gift the work of another man already ready to be patented, but not yet divulged to the general public, Williams felt confident that he now held the world's tail in his right hand and a sharp ox-goad in his left. And how he would make the brute sweat and plod for him when once he started cultivating his rich opportunities! Provided the genius who had invented this irresistible goad had not yet filed the necessary papers, Williams cared not a damn for any moral rights the man might have in his masterpiece; the legal technicalities alone troubled him. Could he beat the cracked genius to the patent office before the idiot repented of his insane generosity for "purposes purely humanitarian and educational"? What wouldn't the rejuvenated P.T.C. do to the blustering, overbearing Erickson with this pointed stick in its capable hand?

From gloating over his anticipated revenge on his unscrupulous rivals, Williams, gazing absently over the fleeting housetops from his seat in the passenger plane, soon fell to speculating on his faithful associates and superiors at the flattened P.T.C. Who among them all had greatly concerned himself with the Vice President's comparative ruin? Not one; their only concern was to salvage at least the rind of their own bacon from the general mess. He, they intimated, was no longer useful to them. Therefore he might go to the devil as fast as he liked. Williams began to smile. His friends were no longer of interest to him. Could they raise capital to finance the goad? They could not. He, on the other hand, with an argument like this patent—which he now felt sure of obtaining—could persuade all the bulls and bears in Wall Street to dance jigs on their heads for his pleasure.

Williams, in short, was not one of those rare souls whom prosperity does not corrode. Had he but guessed that his anonymous benefactor intended by his gift that the recipient *should* go to the devil, as the P.T.C. had already hinted, his smile might have been less confident. The joke after all might turn out to be on him, as a mere pawn in the humanitarian and educational purpose of the donor.

It must have been the very morning that Williams rushed off to the U. S. Patent Office that De Soto rose much earlier than usual and, while Alice still slept, stole out to the back garden to inspect his menagerie of pets. For the past three or four days one of the guinea pigs and both of the frogs which he had taken to the laboratory had been acting strangely. In no case were their actions those of animals in normal distress. Each seemed to sense in some mysterious way the nature of the unseemly jest which chance—or design—had played upon it, and each of the hapless creatures appeared to be anticipating with an unnatural dread the miracle which was almost upon it. The natural rhythm of its vital functions had been violated.

Walking slowly over to the cage where the ailing guinea pig lay, De Soto took a firm grip on himself.

"In five seconds now," he thought, with a rueful laugh, "I shall know how long that blundering fool Crane had left the tube running."

A sack had been laid over the top slats of the cage, as the light seemed to irritate the prospective mother. With a firm hand De Soto raised the sack and peered down into the cage. The miracle had happened in the night. In one corner of the cage the wretched mother cowered in unnatural fright, panting with terror. The eyes of the stricken animal, already clouding at the approach of death, were fixed on the farther corner, opposite her own, where lay the four things to which she had given life against her will. De Soto had half expected a shock. But even he was unprepared for what he saw. He replaced the sack, strode to the garage, and fetched a shovel and a bottle of chloroform which he had concealed in the tool cabinet a week before—when the guinea pig first showed signs of distress. In ten minutes he had done what was necessary.

"Now for the other," he muttered, going toward the pen where the suspect frogs lived. Again the miracle happened. In this instance there were no unnatural young. Frogs propagate from eggs. Therein this pair had an advantage over the guinea pig. What their offspring might be was yet undetermined. De Soto decided not to wait for outraged nature to reveal the unknown. The two repulsive monsters, whose grotesquely budded bodies made his blood run cold, were sufficient. Once more he used the chloroform and the shovel.

"I know now how long the tube was running," he thought. Involuntarily he began feeling his muscles and running his fingers lightly over his skin to detect the incipient nodules. "Am I to go like the frogs," he muttered; "or are only my germ cells affected? One or the other; but which? Perhaps both."

He walked slowly back to the garage to put away the shovel. The half bottle of chloroform being of no further use, he intended emptying it and throwing the empty bottle into the rubbish can. Drawing the cork from the bottle he started to pour out the remaining chloroform, and paused irresolutely.

"I wish I knew," he muttered, staring moodily up at the window of his bedroom. His bride of three weeks was still asleep in that room.

Whatever may be a man's abstract theories about humanity as a whole, three weeks of marriage, and especially the first three, will modify them in detail. Moreover, De Soto had undergone a profound physical change the first night of his married life; he was no longer, mentally at any rate, the man whom Alice had married that happy afternoon. Among other discoveries of those three weeks, De Soto learned that he was beginning to love his bride in the human and humane way of ordinary men whom, three weeks before, he had despised. "I wish I knew," he repeated, still undecided. A vivid image of what he had seen in the cage with the dying guinea pig flashed into his mind. Hesitating no longer, he recorked the bottle, slipped it under his coat, and stole into the house. In the kitchen he selected a clean dish towel and stuffed it into a coat pocket. Then he crept upstairs.

Alice was still sleeping, her bare arms gracefully disposed on the silken sheet, and her ruddy lips slightly parted like a child's. She was smiling in her sleep as De Soto stealthily extracted the cork from the chloroform bottle and drew the dish towel from his pocket. For perhaps five seconds De Soto stood motionless, staring at her beautiful face with something like dawning compassion in his eyes. Then he began pouring the chloroform, a few drops at a time, upon the towel. As the sweet, sickly odor flowed slowly down on the rosy face, the sleeper stirred slightly and murmured a word that sounded like her husband's name. Although his hand shook, De Soto did not desist.

At last the towel was saturated, and De Soto laid the bottle noiselessly on the floor. He straightened up, his muscles stiffened for the inevitable struggle.

"Where are those flowers?" the sleeper murmured, now half awake. "Miguel!" Her eyes opened fully, if drowsily. Instantly he thrust the towel under his coat. "What is it?" she asked, starting up. "You got up early?"

"Yes," he replied slowly. "I thought I smelt gas escaping from the refrigerating plant. Do you get it?"

"I thought I was dreaming of acres of red roses. Now that you mention it, I do notice a sweetish smell. Have you been downstairs to look?"

"No, I was just going when you woke up. Don't worry; I'm sure it's nothing serious. I'll open the dressing-room doors and let it blow out. Where are my slippers? Oh, here they are."

Bending down quickly he managed to secrete the bottle under his coat while pretending to put on his slippers.

"Hadn't you better call the servants?" she suggested as he flung open the doors of the dressing room.

"No," he laughed. "It's still nearly three quarters of an hour ahead of their usual time. You forget that I'm a sort of glorified tinker myself. I'll soon fix whatever is wrong. Now you take another nap; I may as well stay up now that I'm dressed."

"If she hadn't opened her eyes," he muttered as he descended the back stairs to the kitchen, "I could have done it. Now I never can. There must be some other way of neutralizing it in her—if it has happened. Why can't I think clearly as I used to think? Well, I can only try. This blind fighting in the dark—"

Although De Soto did not know it, and indeed was incapable

of realizing the fact, he not only was changed but was also in the merciless grip of a slow but incessant transformation. He was like a robust man of splendid intellect suddenly assailed by an insidious and incurable disease of the bodily functions and mental faculties. Such a man, in the first, gradual stages of his decay, perceives nothing wrong with himself, and attributes his slackening grip to an inexplicable conspiracy of outward circumstances. The problems he could have attacked and solved in his prime baffle him in his decline because—according to his rationalization of his disease—they are more abstruse than any to which he is accustomed. His friends, pitying him in his decay, do not disillusion him, and he goes to his grave believing that the world has passed him in its ceaseless progress, whereas it is his own rapid retrogression that has shot the world ahead beyond his ken.

The first three weeks of Alice's honeymoon had passed in a happy dream, at least for her. Every morning she accompanied her husband to his laboratory and passed the day pretending to read but acutally following his every movement with devoted eyes. She proved herself an ideal companion for a desperately busy man, talking only when he showed an inclination for talk. When lunch time passed unobserved by her husband, she would slip out to the nearest restaurant, to return presently with an appetizing meal, which she spread out unobtrusively where he might notice it.

Even in the first week Alice observed a curious change in the man she worshipped. Mistakenly, she imagined that he was working too hard. All the staff had told her such wonderful tales of the lightning sureness of his mind, that it puzzled her to see him frequently baffled. Unaware that she was watching his slightest movement, De Soto would often sit for minutes at a time, turning some piece of apparatus over and over in his hands, as if in doubt concerning its use, although he had made it himself. These lapses became more frequent as the construction of the new tube progressed, until by the end of the third week practically half a day would be wasted in futile scribbling or blundersome manipulations. Alice became alarmed, and begged him to take a rest, if only for a week.

His reply was a stare of unfeigned surprise. Wasn't he getting along famously? Why interrupt the work with the end in sight? With a chill feeling about her heart, Alice realized that her husband was headed straight for a nervous breakdown, and was so far gone that he failed to appreciate his illness.

At length Alice could stand the suspense no longer. On the morning of the day when De Soto had been tempted to destroy her, she asked his permission to invite an old friend to dinner that evening.

"Of course," he agreed readily. "Who is it?"

"Doctor Brown. I haven't seen him since I was married, and he was so good to father and me."

"Why not invite your father, too? It must be pretty lonely living at a hotel."

She hugged him in an ecstasy of happiness. De Soto, for his part, felt an unaccustomed uneasiness at the prospect of a meeting with Brown. Would the doctor inadvertently refer to the strange disorder of which De Soto had never told his wife? He must see Brown first and warn him to be silent. Then a disturbing question echoed through his mind: Why must Brown be warned? Surely there was nothing disgraceful in a man keeping a passing sickness from his wife? Ah; De Soto remembered —but not clearly. It mattered nothing whether Alice learned of his itching skin, now permanently cured. No; but she must never hear of what happened in the laboratory that night when he caught Crane trying to work his tube. A worried frown darkened his face. Exactly what had happened that night? The main events stood out fairly clearly, but the details were blurred almost beyond recall.

"Alice," he said, "I guess I'll take a layoff next week, after I finish my tube. It will be done this morning, I hope."

"Oh, how jolly! That's just what I've wanted you to do ever since we were married. Only," she added in a low voice, her eyes shining with unshed tears, "you seemed to think so much of your work that I never dared to hint."

For some minutes he remained coldly silent. Had she displeased him by her outburst of affection? Alice glanced shyly at his face. Why was he so withdrawn into himself, so far away, in seeming, from her and the world she knew? At length he spoke, more to himself than to her.

"Something happened a long time ago, when, I can't recall. But it was so far away in time that it seems like a dream from another life. What was it? Why can't I remember? And why should I always seem to be on the point of meeting someone whose existence I have forgotten?"

"Never mind," she said soothingly; "if it is anything that you should remember, it will all come back after you have had a real rest."

As De Soto had prophesied, the new tube was finished and ready for operation that morning. Shortly before noon it was connected to the twenty million volt box, all but the last terminal which would close the circuit and start the generation of the rays. Alice at the time was pretending to read. Apparently she was absorbed in her story. De Soto furtively studied her profile a full minute and then went to the closet where the insulating suits were stored. Presently he emerged, clad from head to foot in a double sheath of the transparent armor. Alice put aside her book and laughed.

"How funny you look in those things! I never saw you dressed up that way before. What's it for?"

"Oh, just for a fussy precaution," he replied lightly. "You see, there might be a faulty connection that would cause a spark. This makes everything perfectly safe, no matter if the whole box blows up. But it won't, so you needn't worry. You stay over there."

He did not act in haste. As dispassionately as he could, he weighed the probable consequences of what he was about to do. His penetrating insight into the laws of nature was already clouding. Like the ordinary man of genius he was now reduced to weighing probabilities and selecting what appeared the least undesirable. Involuntarily shutting his eyes he quickly turned on the full twenty million volts for an instant, and then off again, by two quick twists of the screw switch.

The shriek that Alice emitted sounded scarcely human. Although De Soto had expected it, his blood froze. Tearing off his hood he ran to her. She had not fainted, but stood staring at him like a shadow in a dream, her eyes dark with terror.

"What was it?" he cried, as if he did not know.

"Are you hurt?" she gasped.

"No. Don't you see? Why did you scream like that?"

"I don't know. For a second I thought you were killed. Then something seemed to tear me to pieces—inside, here."

"Imagination," he boldly reassured her. "You feel all right now?"

"I suppose so," she admitted doubtfully. "But I feel—oh, how can I express it? Changed." Then, after a pause she added in a voice which he scarcely heard,

"Defiled and degraded."

"Nerves, Alice. You imagined that what I was doing was terribly dangerous. Sorry I stirred you up by putting on all this ridiculous fancy dress. When I turned on the current you thought I was killed. Come on, let's go out to lunch and get some fresh air."

Still dazed, she sat down and waited until he removed his protective armor and put on his coat. What had happened to her? Merely an attack of nerves, probably, as he asserted. Yet she felt inhumanly unclean. By a curious coincidence the warning which

Crane had thrust upon her recurred now with startling clarity. As if her old friend were standing before her, she saw his face with her mind's eye and heard his disturbing prophecy: "If you marry him you will kill yourself to be rid of him." What did Crane know of her husband that she did not? Surely nothing, she concluded reassuringly as her common sense regained the control of her subconscious mind. Then, from her deepest nature, just as her husband reappeared, a despairing instinct whispered, "Destroy yourself before it is too late." But this warning, like Crane's, yielded to gross common sense, and she joined her husband with a smile.

"I'm all right now," she said, and believed it.

On the way to lunch she telephoned to her father and Doctor Brown, inviting them to dinner that evening. Both accepted eagerly, especially Kent, who was longing for a sight of his daughter. Brown looked forward to the evening with mixed feelings. To sit down at the dinner table with a man whom you distrust and whom you have wronged is likely to be rather trying.

That evening Kent arrived first, three-quarters of an hour ahead of time. After a decently cordial greeting, De Soto retired to inspect his pets, leaving father and daughter together to discuss him to their hearts' content. To the happy father's uncritical eyes Alice seemed the picture of health and youthful happiness. Kent himself was in high spirits. For a week he had been employed as booster-in-chief for a go-getter real estate firm, and was enjoying his work tremendously. As the time for Brown's arrival drew near, Alice hinted that she would like to see the doctor alone for a few minutes. "About Miguel," she explained. "He has been overworking. Suppose you go out and ask him to show you his pets? He's crazy over them, and will let no one else have any of the care of them." The bell rang just as Kent made his escape.

"Well, Alice," the doctor greeted her, "this is just like old times. How is everything with you?"

"I'm ridiculously happy," she laughed, "except for one thing. Won't you drop Miguel a hint that he must take a long rest?"

"From what the men in the laboratory tell me, your husband isn't given to long rests. Still, I shall do my best, if you wish it. What seems to be the trouble?"

"First let me tell you that I spend my days in his laboratory, reading and watching him work. He is usually so absorbed that he doesn't know I'm there. So I can't help seeing him as he really is. And I have noticed that he is dreadfully tired, although he does not know it. For one thing, he has long lapses of memory."

"I'll speak to him," Brown replied decisively. "We can't afford to have him unwell or you unhappy. What about yourself? Feeling pretty fine?"

The doctor eyeing her keenly, noted the slight hesitation and the flush before she replied.

"Never felt better in my life," she began. An overpowering wish to confide in her friend suddenly stopped her. Almost before she realized what she was doing, she had told him of the excruciating momentary agony she had experienced that morning in the laboratory. "It was probably just an attack of nerves, wasn't it?"

"Tell me exactly what happened."

She went into detail, describing the whole incident and De Soto's explanation. Brown, of course, noticed the flaming fact that Alice was unprotected while her husband neglected no precaution to shield himself against possible danger. Like a good doctor, Brown's face betrayed no concern. Nevertheless he was revolving in his mind a black question that chilled him to the bone. Had De Soto merely blundered, or had he intentionally left Alice unprotected? And if the latter, what could be his object? For Brown vividly remembered certain cries which he and Crane had heard—those of a helpless animal. What would a human being suffer in similar circumstances? Had Alice indeed been subjected to the same treatment as Bertha?

"Miguel was right, wasn't he?" she concluded. "There couldn't have been a leaky connection, or we should both have been killed. It was just my nerves."

"Not a doubt of it. You take my advice and keep out of the laboratory after this. The next time something real might happen. By the way, you have no tingling or itching of the skin?"

"Not a trace."

"Then that settles it," the doctor assured her. From his tone she inferred that he dismissed the flash of pain as a fiction of her imagination. That of course was precisely what Brown meant her to believe. The point, however, which her healthy skin settled in the doctor's mind was more important. She had taken the full bolt of the rays generated by twenty million volts, if indeed she had been exposed to any, and not the greatly softer rays given off at two million volts' pressure. Kent and De Soto joined them just as dinner was announced.

During the meal Brown concentrated his attention on De Soto, leaving Kent and Alice to gossip of old times at the Foundation. Poor Kent, in spite of his pride in his new job, longed for the fleshpots of his lost dictatorship. A jealous note crept into his voice, and the doctor overheard him surreptitiously expressing a hopeful belief that the Erickson would come to a bad end. They were too grasping, he declared, caring nothing for the common decencies of reputable business competition. De Soto overheard the remark.

"I agree," he said quietly. "The trustees need educating."

"In what?" Brown asked.

"Human decency, if there is such a thing."

"Miguel!" Alice murmured reproachfully. "You know you don't mean that."

"No, I meant more." His voice rose. "The whole human race needs educating—in the same way. Why, I remember when I was a young man—."

"You can't be so very ancient now," Brown interrupted, with a curious glance at his host's excited face.

It was an unfortunate remark. Something snapped in De Soto's brain. Flinging down his napkin, he pushed back his chair and leapt to his feet, his black eyes blazing. Luckily no servant was present at the moment. Speaking with great rapidity and in a low voice vibrant with passion he delivered a flaming tirade against everything human. Alice watched his face with something akin to terror in her eyes; Kent sat open-mouthed and blank; Brown followed every word with rigid attention. An alienist, knowing nothing of the facts, would have pronounced De Soto incurably insane. The very logic of his fantastic indictment was its most damning feature. Brown was not an alienist. But he had the average high grade physicians' knowledge of the earmarks of insanity.

Silently admitting to himself that any specialist on mental disorders would be fully justified in declaring De Soto insane, he nevertheless felt confident that the raving man was sane with a terrible sanity denied most human beings. The outburst lasted but a brief two minutes. It was like a terrific stab of lightning on a sultry midsummer night. Breathing heavily, De Soto resumed his seat and began crumbling a piece of bread.

Brown broke the sulphurous silence. With a significant glance at Alice, which he interpreted correctly by kicking her father under the table, the doctor began a cool cross-examination of De Soto.

"Your theories of human society are interesting but academic. How can you put them into action?"

"Oh," De Soto laughed, apparently himself once more, "I

can't. My theories are just theories, nothing more. I thought they might amuse you."

"They did. You seriously think it would be possible to educate human beings out of their greed for what you call trash by stuffing them with so much of it that they would rebel?"

"Not exactly. That would be merely the first step."

"And the second?"

"Give them different tastes. Even that cook and waiter had rudimentary minds that the right process could work on."

"What cook and waiter?" Brown demanded quietly.

"I just told you."

Before continuing his examination, Brown shot Alice a warning glance.

"Of course," he said. "I forgot. Let us suppose you have made the rest of us disgusted with the things we like. Would you give us something bettter?"

"I was going to. But—"

"But what?"

"Oh, what's the use of theorizing? Leave it to history."

Brown changed his tactics. One or two statements of fact which De Soto had let fall in his tirade needed attention.

"As you say, history will attend to our descendants—unless we find some way of doing it ourselves. Another thing you said is more interesting, I imagine, to all of us. You have always been rather a mystery man to most of us, Mr. De Soto. We never knew that you spent some of your earlier years in the United States."

"Neither did I," De Soto retorted with an amazed stare. "Who said I did?"

"My mistake," Brown apologized. "But it seemed to me that the conditions you mentioned could exist only in the United States."

"They exist everywhere."

"In the Argentine, for instance?"

"The Argentine?" the puzzled bewilderment on De Soto's face showed plainly that he did not perceive the drift of Brown's question.

"I just used it as an example of 'everywhere'," Brown explained.

It was clear that De Soto either was lying or that he was so ill that he remembered nothing of his early life. Kent was about to break in when Alice silenced him with a warning look. She, too, had believed that De Soto spent his youth in Buenos Aires. Poor Miguel was indeed unwell. She hurriedly turned the conversation into less personal channels.

The distressing party broke up early. At a hint from Alice, Kent left immediately after dinner, saying he had to be up very early to keep a distant engagement. Alice followed him to the door, leaving Brown alone with her husband. She did not hurry back. The doctor caught her expressive glance as she went out.

"Mr. De Soto," he began as soon as they were alone, "you are too valuable to society to overtax your strength the way you do. If you are not to squander all your talents in a silly nervous breakdown, you had better take a long rest."

"I am planning to," he replied. "My work is practically done."

"Not all of it, surely?" the doctor suggested. "That's no state of mind for a young fellow like you."

De Soto flared up again.

"What do you know of my age?"

"Keep cool. Don't fly into rages over trifles. As a matter of fact I don't know your age, but I should guess it to be about thirty."

"Thirty?" De Soto echoed in astonishment. "Why I was born—."

"When?" the doctor prompted. Receiving only a puzzled look, he continued. "You have forgotten that, too. Your wife is right. Take a lay-off."

Alice reentered, and Brown took his departure. She saw him into the hall.

"Your husband will be all right," he reassured her, "if he lets up a bit. If there is anything I can do for you at any time, please let me know."

The doctor walked thoughtfully home, wondering what sort of a man De Soto was at bottom. Was Crane right in his estimate, and if so, in what particular way was De Soto a thoroughly bad egg? Brown half doubted his friend's opinion after seeing the suspect in action. More likely the brilliant young inventor was merely eccentric. But was he also a bungler in practical details? That was the hard question. Its answer would decide whether De Soto was guilty or not guilty in regard to the mishap to Alice.

As he passed along the south wall of his garden, Brown heard a prodigious fuss from the hens in the patio. At this hour of the evening they should all have been roosting and silent. That they were active and excited in the dark, in flagrant contradiction of the normal habits of the fowls, presaged some event of unnatural significance.

"Bertha's laid another," he exclaimed, hurrying into the house to fetch his flashlight.

Since her involuntary adventure in the laboratory, Bertha had set out to beat the world's record in laying eggs. She had already broken the record in the matter of numbers. The size, however, of her efforts disqualified her. None of her numerous eggs had been over a fifth the size of a normal hen's egg. The shells, too, of these "pigeon eggs," as Brown called them, were remarkably deficient in lime. Some were little more than sacs of flexible white skin, like the inner sheath of an ordinary egg. Naturally the doctor had watched his phenomenal hen as closely as if she were his wealthiest patient. She had laid no fewer than sixty-three of the dwarf eggs. She, herself, seemed quite satisfied, as she sat almost constantly on the whole nestful.

The henyard was in a wild commotion. The doctor's flashlight revealed an excited dozen or so hens pecking viciously at some dark red object. This proved to be Bertha, her feathers drenched in her own blood. She was dead, but still warm. Brown shooed the enraged hens away from the body and placed it under an empty coop. Then he investigated.

From the evidence it appeared that Bertha had died defending her brood. Eighteen of the eggs had hatched. Thinking for a moment that he had gone insane, the doctor stared down at the dead hen's living offspring writhing over the unhatched eggs. Then he scooped the lot, eggs and offspring, into his hat, hurried back to the house and telephoned to Wilkes and Crane.

"Come over at once. I have some things millions of years old to show you, and they're alive."

He turned one of the crawlers over on its back with a pencil.

"No wonder the other hens pecked her to death. I would have done the same in their case."

10

CAT AND MOUSE

IN SCIENTIFIC CIRCLES THERE ARE SEVERAL semi-human periodicals which contain, in addition to technical papers, brief personal notes concerning the scientists themselves. For example, if Professor X. is appointed to the vacancy created by the death or resignation of Doctor Y., the fact is stated, so that the scientific friends of Professor X. may know where to address him.

Crane, of course, took advantage of these free employment agencies when De Soto discharged him, and sent in a note to each, saying that he was no longer connected with the Erickson Foundation. He hoped to receive at least one tempting offer before his bonus ran out. The hope was not extravagant. Before De Soto's brilliance had eclipsed that of all inventors in his own many fields combined, Crane was justly rated as the best ray expert in the country. Hence, should some desperate firm attempt to hold its own small corner of the field against the Erickson, Crane was their most promising prospect, as De Soto seemed satisfied where he was. In fact several firms were already considering Crane when the official news of his "resignation" was published. They knew that his nose was out of joint at the Erickson and hoped to get him cheap. In this they were disappointed. Other attractions held Crane jobless to his post.

At last, five weeks after the evening when Wilkes and he responded to Brown's excited telephone call, Crane received a three hundred word telegram, signed Andrew Williams, President Universal Power Transmission Company, offering him a royal salary as chief consulting physicist. The telegram, while avoiding all details of technical value, stated that the new company had been formed to exploit a revolutionary invention for the transmission of electric energy. This much occupied less than twenty words. The remainder of the telegram was chiefly a roll call of the wealthiest business men in America. These, Williams stated, were floating the company on their own money. That list of names would have impressed the Sphynx.

"Hang it," Crane muttered, "I can't turn down an offer like this and keep my self-respect. I shall have to accept, just as Wilkes and Brown are getting to the most exciting point."

Before telegraphing his acceptance, he called on the president of the Erickson trustees. This time the president bustled out in person to greet his caller.

"Come into my office," he begged. "Well," he asked when they were alone, "you have found something about the subject which we talked of two months ago?"

"Nothing of commercial value," Crane admitted.

"Your opinion of Mr. De Soto is the same?"

"Yes. Only more so. I can't tell you why. But I am beginning to get a definite line on him. By the way, did he ever tell you that I spent an hour in his laboratory without his consent?" The president shook his head. "I thought he wouldn't. Well, what I came about is this." He handed the president the telegram from Williams. "It does not say confidential," he remarked; "so there is no harm in your seeing it."

The president read it through slowly twice. Its commercial implications for the Erickson were obvious.

"You will accept, of course?" said the president. Crane pointed out that he would be a fool to refuse. "I agree," said the president. "You will not forget us, I hope? We were not ungenerous to you."

"What can I do that your own staff can't? This invention must have been patented before the company was formed. So anyone can find out by going to the Patent Office exactly what it is. Why not send De Soto to Washington at once? Even if the patents are ironclad to the ordinary man, he will find a way through them."

"Do you mind if I show him this telegram?"

"I guess it's ethical enough. Go ahead."

De Soto was not in his laboratory. On telephoning to his house, the president learned from Alice that her husband was not at home.

"Can you tell me where I could get in touch with him, Mrs. De Soto? This is a most urgent matter; otherwise I should not dream of troubling you."

"He had an appointment with Doctor Brown this morning at eleven. Probably he is at the doctor's office now."

"Not unwell, I trust?"

"Oh, no. It was just about a personal matter."

De Soto was located at Brown's office. He promised to be at the Foundation within half an hour. While waiting for him to appear, the president kept delicately reminding Crane of the great debt of gratitude, which he, Crane, as an altruistic scientist, must feel that he owed the Erickson Foundation. Crane had difficulty in smothering his grins. The game was too obvious.

When De Soto entered, he nodded curtly to Crane, and proceeded to business.

"What is it?" he demanded.

"Doctor Crane advises that I show you this telegram, Mr. De Soto."

De Soto was almost his "old" self for a few seconds. He

took in the sense of the long telegram at a glance.

"Electrical energy?" he questioned with a short, contemptuous laugh. "So that's the sort of thing these great financiers gamble on, is it? Serve them right if they get cleaned out. I have neither sympathy nor patience with them."

"But," the president expostulated, "all of our own business is built up on electrical energy in some form or another. What is to become of our insulation if these people have something that beats it? And our radio valves—everything we manufacture. Don't you see how serious this may be for us?"

"No. It doesn't matter what they do."

"You know what they have?"

"I do not," De Soto snapped. "And what is more, I don't care. If I did wish to learn, I should telegraph at once to Washington for a copy of the patents."

"Hadn't I better do so?"

"Why? I can beat anything they do."

"Really, Mr. De Soto," the president demurred, "although we all have the utmost confidence in your genius, I must say that your attitude strikes me as a little too—how shall I say it?"

"Call it cocksure, if you like. I shan't mind, because I do know what I am talking about."

"But consider this list of names for a minute. Would men of such standing in the business world put their own money into a scheme that wasn't gilt-edged?"

"They would, and they have, because they are one and all uneducated fools."

"I must protest! These men—"

"Do so. And so shall I. In fact I have already protested in the only way that counts. I call a financier uneducated when he puts a lot of his money into a scheme that he does not *know* will win. As I said, I can beat anything they do, and I don't mind if Crane tells them so. They won't believe it."

"But what are we to do?"

"Nothing, for the present. Wait until they are in up to their necks. Then I will finish their education by shoving their heads under. It is either this or nothing. If they are right, we are ruined; if I am right, they are ruined. We can't compete with them on their terms—if they have what they think they have. And neither can they compete with me, if I have what I think I have. Our policy is plain—wait."

"May I suggest a third possibility?" Crane interjected, as the president was about to reply. "Mr. De Soto says either they or we shall be ruined. We might properly consider the case in which neither is ruined."

"Impossible," De Soto snapped.

"All right," Crane retorted. "There is only the fourth thing possible."

"There is no fourth," the president objected.

"Oh, yes, there is. There are four possibilities, and only four. We've discussed three. The fourth is the least pleasant of the lot. Both we and the other crowd might be ruined."

"But how?" the president demanded, missing the ominous flash of De Soto's eyes which Crane observed. "Industry must have electrical energy. How can they lose, provided we also lose?"

"I'm sure I don't know," Crane admitted. "I'm not a great inventor."

And there the matter rested. Forced to accede to De Soto's policy of inaction because he could devise no better, the president delivered the fortunes of the Foundation into its director's capable hands. Crane left them arguing and went out to wire his acceptance. His sympathies were with the president; De Soto seemed entirely too sure of himself for comfort.

Before taking the train that night, Crane went to bid the doctor goodbye.

"I hate to rush off and leave you and Wilkes just as things

promise to get exciting. But what can I do? Offers like this don't come in every mail."

"It's the only sensible thing to do," Brown agreed heartily. "I'm glad you had time to drop in. There is a new development." His face darkened. "Professionally, I have no right to tell you. But in this case we are beyond ethics. I shall also tell Wilkes. We three are the doctors in this case. What I tell you must go no farther."

"I promise."

"Alice called me in this morning. She is to have a child."

"Good God! What will you do?"

"Take care of her, of course. She asked me to."

"But—"

"I know. Or at least I don't know. Neither does her husband. He came to see me this morning, after I got back from visiting her. I believe he has blundered and tried to correct his mistake. But he is not sure of anything."

"Did he tell you?"

"No. It would be impossible for him to commit himself beyond the vaguest suggestion."

"Of course. What did he say?"

"Nothing true or of any unmistakable consequence to us. He merely insinuated that Alice's health is not good enough to stand the strain. Having just examined her, I knew that he was lying, and I suspected him of wishing me to know that he was lying in what he considered a good cause. You see, of course, he could give no hint of what may be in his mind. As I said, I feel that he himself is not sure of his ground. Still he does seem to suspect that he may have failed. What could I do? I was bound, professionally if for no other reason, to ignore his suggestion."

"What if he consults some other physician?"

"How can he? Any physician who would do what he hints —and there are plenty, I admit—would not be safe. A reputable man is the only one who can be trusted in a case like this."

"But there hasn't been another case like it, if—"

"Not exactly; that is true. But there have been several on the same level ethically. You see what I mean."

Crane brooded miserably in silence for some moments.

"I wish she would die," he said at length.

"So do I," Brown rejoined. "But there's no hope. She's too healthy."

Six days later, in New York, Crane and his new employers held their first conference over the epoch-making invention which, according to President Williams, was to revolutionize all industry. From the moment he set eyes on Williams, Crane disliked him. The suave president of the new Universal Power Transmission Company was inclined to be fleshy, although still in his forties, with a perpetual and exasperating smile of self-satisfaction greasing his smug features. His assumed joviality and mock good-fellowship made Crane long to smash him in the face.

"What do you think of it, Doctor Crane?" Williams beamed. The patent papers and a rough, small scale model lay before them.

"It's a washout," Crane admitted. "By the way, you haven't told me the inventor's name."

Williams laughed obesely. "I thought you physicists were keen observers. The name is plastered all over the 'Evidence of Conception' alone, to say nothing of the final patents."

"So I observed," Crane remarked dryly. "But you haven't answered my question. Who invented the thing?"

There was an ominous silence. Williams' face lost its oily joviality as he glared at Crane, and the consultants of the staff fixed their chief with doubtful, questioning eyes.

"Do you mean to insinuate," Williams demanded in a voice that cut like steel, "that I did not invent this method of power transmission?"

"Not at all," Crane responded promptly. His tone was conciliatory. Instantly it changed. "What I mean is this. You are a . . . liar and a thief."

Williams leapt to his feet, trembling with rage.

"What do you mean?"

"Exactly what I said. You did not invent this. I have known of your work for years—ever since I was a sophomore at the University. It was good stuff—no doubt of it. But the best you ever did was not within a million miles of this." He paused, to emphasize his point. "Gentlemen," he said, addressing the staff, "it is as impossible for President Williams to have made this invention as it would have been for the village idiot of Stratford-on-Avon to write 'Hamlet.' You see the point? President Williams' own stuff is several thousand levels lower. Therefore, I say, he stole this invention."

The silence grew oppressive. Williams broke it.

"You may leave me to settle with Doctor Crane," he said to the silent group. "This is a personal matter. Unless," he added, addressing Crane, "you prefer to apologize publicly?"

Crane shook his head and the staff filed out.

"Now," Williams began when they were alone. "You will withdraw what you said."

"How do you know? As a matter of fact I shan't. I don't want your job.

"You have no job, Doctor Crane. I was not offering you a chance to retract and be taken on again. It is now merely a question of whether you wish to stand suit for libel. The staff heard what you said."

"I'll say it again, if you'll call them in. And I'll tell them who made that invention, if they care to know."

In spite of an effort to control himself, Williams went the color of a dead cod.

"Who do you think made it, if I did not?"

"Miguel De Soto. It has all the earmarks of some work

he has been busy on since he joined the Erickson. I would recognize it in my sleep. How did you get hold of it?"

Without a word, Williams rose and opened his private safe.

"You know too damned much," he admitted with a cynical laugh, thrusting the thirty-page anonymous letter into Crane's hands. "What do I care who knows where I got the stuff? It's mine, and I have the patents."

Crane read the letter through. The only point of interest to him was the postscript: "P.S. *My purposes are purely humanitarian and educational.*"

"Well?" Williams demanded as Crane handed back the letter.

"I knew it. De Soto invented your wireless power transmission. The postscript would give him away at once to the President of our Board of Trustees. De Soto is a bigger fool than I thought he was. He has blundered again."

"How?"

"By mentioning education. It's too long a story to hash over again. Besides, what is there in it for me?"

"If you know any facts of value, I could see that you are well paid."

"All right, I know one fact that will save you hundreds of millions—possibly a billion or two."

"Does it concern the invention?"

"Vitally."

"How much do you want for it?"

"One hundred thousand dollars, paid in advance, in thousand dollar bills—common currency. No stopped checks for me."

"You can be insulting when you're in the mood, can't you?"

"I haven't tried yet, so I don't know. Take my offer?"

"I'll consider it if you tell me why De Soto sent me that letter."

"Because he knew you would bite."

"Then you consider the invention deficient in some detail we have overlooked?"

"Not at all. It will work."

"Where is the catch, then?" Williams demanded.

"I don't know."

"Yet you have a suspicion?" Crane nodded. "Very well," Williams concluded, "I'll give you a hundred thousand for your fact. What is it?"

"Easy. Write out this: 'I hereby pay to Doctor Andrew Crane, for technical services rendered, one hundred thousand dollars in U. S. currency, thousand dollar denomination, numbers' —leave space to write in a hundred long numbers. Then you come down to your bank with me and get the bills. We'll have your signature witnessed by the cashier and a couple of clerks, and the numbers of the bills written in. Then you can't stop anything on me—unless you hire a gunman."

"You must think we're crooks," Williams retorted coldly. Nevertheless he wrote. "My backers have billions," he added with a touch of snobbery. "A hundred thousand for vital information is not an unreasonable fee."

"You bet it isn't. Ready?"

At the bank Crane refused to divulge his 'fact' until the bills were safely in his pocket with the duly witnessed statement.

"Now," Williams demanded when they reached the street, "what's your tip? You have your money."

"Just this. Throw away the invention and dissolve your company."

Williams glared at the lanky young man before him in speechless rage.

"You—" he sputtered.

"Keep cool. That tip is worth all the money your crowd has. If you touch the invention, De Soto will break you."

"How do you know?"

"Because he told our president so just before I left. I heard him. Of course he did not tell us that he had made a free gift of this invention to you. Some things are better left to the imagination."

"But why—"

"Because he hates your methods of doing business."

"What about your own?" Williams flashed.

"The same there. He doesn't like any of us. My idea is that he plans to break all of your crowd first and attend to us later. What he doesn't know about the business mind isn't worth knowing. He knows that all the big money in America would fall for a sure thing—it certainly looks sure enough—like the wireless transmission of all electrical energy at a tenth of a per cent. of what transmission costs now. Who wouldn't? I'd have fallen for it myself, if I hadn't known De Soto. Can't you see? It's all so simple. Your crowd puts all its cash into a sure bet and finds out the day after tomorrow that there is no market for what it sells. Where are you? In the soup. It will cost money—lots of it—to manufacture this device on a world-beating scale, as you intend. Go to it; De Soto will bankrupt you the day you begin to market."

"If I thought you knew what you were talking about," Williams muttered, "I would call it off now. We've already spent four hundred and fifty million in buying up strategic locations for our plants."

"Better swallow your loss and back out. You'll be smashed. Call it off. De Soto is a hardboiled Tomcat and you're an innocent little mouse."

Williams was one of those high-powered Captains of Finance who made lightning decisions in a bold, impressive way, and frequently kicked himself afterward for the heady, natural fool he was. His square, beefy jaw set.

"I'll see it through," he decided, as if he were Napoleon at Waterloo. "This can't be beaten. You're welcome to your fee."

"Thanks. And you are more than welcome to my tip. If

I can be of further service, here's my address. I'm going back tonight."

The game was now becoming fairly clear, especially to Crane, who knew certain facts not yet divulged to the commercial world. Feeling that his loyalty—if he had any—was still to the Erickson, Crane did not wait for the slow transcontinental railway service to get him home, but engaged passage on the combined express and passenger plane routes. Twenty hours later he was in Seattle, telephoning to the president of the Erickson Trustees.

"I'll be at your room in fifteen minutes," the president promised. And he was.

Crane's report was disturbing enough. What could be De Soto's object?

"I can't understand him," the president admitted after a two-hour session during which every aspect of the singular situation was minutely examined. "Well, I can soon put him to a test. If he is on the square, and really has our interests at heart, he will tell us at once how we can beat the Williams crowd."

"He hasn't your interests at heart," Crane remarked quietly. "As I told you the morning he fired me, he hates your guts."

"I'm not so sure," the president demurred. "De Soto is a genius, in business as well as in invention. How do you know but that this scheme of his to trap all the big fellows isn't just a fine evidence of his loyalty to us? After all we have treated him handsomely. What more could he ask? We've deferred to his slightest suggestions. Who made him rich? We did. No; I believe De Soto will make good and show us how to break the Williams crowd flat."

"The biggest smash in the history of American big business," Crane mused. "When it comes, let me know. I'm putting my pennies in a safety deposit vault till De Soto is shot. Now," he continued with his slow grin, "if I really loved humanity I would go out and shoot De Soto now, instead of waiting six months or a year for some busted banker to finish the job as it should be finished. I know a lot about our friend that you don't."

"What?" the president demanded, going white in a vague panic.

"I can't say yet. The information isn't mine to give out."

"Then who can say?"

"That would be telling. I've said my say. My advice to you is the same as that I gave Williams. Fire De Soto, shoot him, have him locked up—anything you like—but get from under him at once. Otherwise he will explode and blow you and all your crowd into little bits. Do I get anything out of this?"

The president reached for his cheque book.

"If Williams could afford to fee you, I guess we can. We're not paupers yet. I shall watch De Soto."

Crane nonchalantly glanced at the cheque.

"Thanks," he said, concealing his elation. "Take my advice and get out. Let the Williams crowd swallow the loss."

The president decided to take the devil by the horns immediately.

"Would you care to come with me and repeat your story before De Soto?"

"Not in the least. Where is he?"

"At home, resting. Mrs. De Soto is not very well, and neither is he, I imagine."

"All right, I'm game, provided we don't run into Mrs. De Soto," Crane agreed. "The last time I called on her she was Alice Kent. She showed me the door. So make this strictly a business call."

They drove to the house and were admitted at once. Alice did not appear to greet them. It was a full ten minutes before De Soto entered the reception room. When he did, a strong odor of chloroform accompanied him.

"One of my pets was suffering," he explained, seeing that

they noticed the smell. "Excuse me for having kept you waiting. Mercy first," he concluded with a strange smile.

"Pardon me for coming to your house on a business matter," the president began, "especially as you are not feeling very well. But I thought you would be interested in hearing Doctor Crane's report of what happened in New York."

"I can guess it," De Soto replied indifferently. "You remember that I gave him my permission to tell Williams that I can beat anything his firm does. Crane told him; Williams didn't believe him. Is that how it stands?"

"Exactly," the president nodded.

"And you wish me to make good on my brag—as you thought it was?"

"It seems to me, Mr. De Soto, that we have no time to lose."

"I agree. Shall I come to your office at three o'clock this afternoon? Very well. Please ask the technical staff to be present. I shall explain to you and the other trustees exactly what I propose to do. To the staff I shall give only the necessary instructions for making full-size instruments for demonstration purposes. The finishing touches must be done by me when the technicians have completed their part—say about eight or nine months from now. Then, with the perfected apparatus in our hands we can get our patents in short order, just as the Williams crowd is beginning to sell. We shall scrap all of their plants and the rest of their investment overnight. To manufacture their device on a world scale—which is what they will do—will take practically all of their capital. They won't be content with the American demand, but will strike from the first for the world market. Let them; so much the better for us. In half an hour, or in one hour at most, I will destroy their world market before they have delivered a single transmitter."

"But how?" the president doubted, his eyes rounding with cupidity.

"Later—eight or nine months from now. Let me have a little fun and I'll give you the world to play with. All I ask is the opportunity, when the time comes, to wreck them utterly in half an hour. I'll make it spectacular," he laughed. "No one shall get hurt—except Williams' hand-picked mob of moneyed easy-marks. They'll be flattened. Financially, only of course. Then you can step in and take the world market they have paid for with their millions of dollars' worth of bribed publicity. If I can't convince you this afternoon you may forget the second."

The president was almost convinced. Still, the fact which Crane had uncovered regarding the origin of the "Williams power transmitter," caused him a twinge of uneasiness. If De Soto could go out of his way to injure men whom he had never seen, what would he do to his daily associates when their backs were turned?

"Is Doctor Crane right," he asked, "in thinking that you sent Williams a thirty-page letter containing the invention they are exploiting?"

De Soto flung back his head and laughed as he had not laughed for months.

"Of course he is right," he chuckled. "But did either of you know that I guessed Crane would see that letter and report what he did? The moment he told me he was going to join Williams' firm, I foresaw everything that has happened—even to this talk and your last question. Aren't you all normal human beings? And don't all such react in the same way to given stiumuli? I couldn't help seeing what has come days before it happened."

"But you wouldn't play a trick like that on us? We gave you your opportunity, remember."

"I shan't forget. Haven't I made good use of your generosity? Here I am offering you the world—that is what your monopoly will amount to—and you look for the trademark to see if it is bogus. Of course it isn't! Wait till I have told you my answer to Williams this afternoon. Then you will see the truth."

"I can't see," the president objected, "why you have gone to all this trouble to deceive Williams if you have something that beats his scheme—your other one, by the way—out of sight. Wouldn't it have been simpler to have started with the winner? Think of the time we shall lose—nine months, you say."

"And you a business man!" De Soto said reproachfully. "With all your possible competitors eliminated before you start, you can gobble up all the markets they might have controlled—wheat, cotton, oil, everything—if you had left them any capital to gamble with. But they will be bankrupt, all their wealth squandered on the one key monopoly they thought they were going to get, but which actually you will have. It is the world I am offering you, I tell you! And you begin to cry because you can't get it for a short eight or nine months. I'm almost disgusted with you," he exclaimed with sudden petulance, which was not at all in jest.

"Don't think me ungrateful or over-suspicious," the president begged. "But as a business man I perhaps see some things more clearly than you, a scientist, possibly can. You say these men will be ruined overnight. Capital can't be destroyed that suddenly. These men are solid—the soundest in America. Their money is not paper. Steamship lines, great banks, whole cities of office buildings, farm lands, timber, and a dozen other tangible things are their actual fortunes. This is no fight on the stock market. We are attacking real assets. Have you thought of that?"

"Yes," De Soto replied wearily. "I know my economics. Also I know my human nature. To build plants to manufacture the new transmitters on a smashing scale, to advertise wherever power is sold or used, to get the sales force into the world field, all of this will require real money by the shipload. And where will our competitors get it? From loans or bonds on all those tangible things you catalogued. Who will lend the money or buy the bonds? Not the big men, as usual, because they are borrowers this time. They will get it from smaller men, little banks, conservative investors, and the great public at large. All these will inherit the big men's office buildings, farm lands, timber and the rest. Then we shall step in and take it all away from them again, for we shall control each and every industry from raw material to ultimate consumer. So much for economics. The human nature of it is even simpler. I needn't explain."

"Perhaps not," Crane agreed. "But would you mind telling us how you got that letter delivered to Williams?"

"Ah," De Soto replied sarcastically, "*there* is a real problem. How would you have solved it?"

"Private messenger, provided I could find one I could trust."

"Good. Just what I did."

From the tone in which he said it, Crane inferred the contempt in De Soto's answer. It seemed to say, 'Here I am offering you the world and you turn aside to fiddle over a trivial problem that an idiot could solve.' The talk was suddenly interrupted by the sound of firm steps descending the stairs. Through the arched doorway Crane saw Brown coming down with his black bag. Excusing himself De Soto hurried out to intercept the doctor.

"Mrs. De Soto is quite ill, I understand," the president confided in a low voice. "Did you notice how nervous and worried De Soto looks? His color is bad, and he has aged ten years in the past week."

"I have noticed the change in color for some time," Crane replied. "When he first came to us he was the color of mahogany, like a full-blooded Mexican. Now he's a sort of lemon yellow, as if he had been living like a beetle under a board for weeks."

"We have tried to make him let up on his work," the president sighed, "but he won't. Some research of his own, he says, has reached the critical stage, and he must carry it through now or lose everything."

"Did he tell you what its nature is?" Crane asked, thinking of the smell of chloroform which followed De Soto wherever he went.

"Something to do with the cosmic rays, I believe, but I'm not sure."

Crane wondered whether the research had anything to do with animals. If so, the strong odor of chloroform would be explained. But he did not share his speculations with the president.

"If you have finished with me," he remarked, "I may as well go. Probably you and De Soto will have private matters to discuss."

"Aren't you coming to the conference this afternoon?"

"I had better not. You see I am no longer officially connected with the Foundation. De Soto might resent my 'spying,' especially as he suspects my feelings toward him. If anything important happens, you can let me know, if you think it wise."

In the hallway De Soto and Brown were conversing in a low tone. Seeing Crane, the doctor stopped short with an exclamation of surprise.

"You back? What happened?"

"Fired as usual," Crane grinned. "Only I fired myself this time. I'll wait for you outside."

When Brown joined him on the sidewalk, Crane briefly summarized his adventures in New York and his conference with De Soto and the president.

"The hundred thousand," he concluded, "with what the president gave me as a tip are enough to make me independent for life. My tastes are rudimentary. Now I can get to the bottom of what friend De Soto has started." He hesitated. "If it isn't a breach of professional etiquette you might tell me how Alice is."

"Everything is apparently normal. If it were an ordinary case I shouldn't have bothered to call. Young husbands are always so fussy in these circumstances that we usually pay no attention to their worries—they mean nothing. But with Alice, of course, I can't afford to take any chances. The thought of what might happen if De Soto called in another physician is appalling."

"You say she is quite normal?"

"Yes," the doctor admitted hesitantly, "except that she is too anxious. There is something not quite natural about her worries."

"I shouldn't wonder," Crane muttered, and changed the subject. "Wilkes is still determined to present his paper at the meeting of the Biological Society?"

"You know how he is. I've been trying to talk him out of it ever since you left, but he insists. Do you want to come along and see the fun? I shall go, of course, if I can get away."

"Sure," Crane exclaimed. "That's why I made Williams fire me. I wouldn't miss that meeting for a million dollars—and I've only got two hundred thousand. It's tomorrow at ten, as scheduled?"

"Ten o'clock, in the university auditorium. Wilkes' paper is first on the program. Well, I must run along to see an old lady with gas. See you tomorrow at the meeting."

"How are your pets coming on?" Crane called after him.

"Too well," the doctor replied grimly. "I've had to build a high concrete wall round my patio."

11

THE TOAD

"THE MAN IS HERE ABOUT THE CHICKENS," THE housekeeper announced the following morning just as Brown was about to begin breakfast.

"What does he want now? I paid him for grain yesterday."

Nevertheless the doctor went out to see what the male harpy sought. Since Bertha's death the doctor had given up keeping chickens on his own premises. Not having the heart to sell his feathered family to the poulterers, he had pensioned them with a farmer in the country. This genius knew a soft thing when he saw it. According to the bills he presented for chicken feed, Brown's pets must have quadrupled their appetites since moving to the country.

The pest extracted a dollar from the doctor—"for grain," he said—but showed no disposition to leave.

"What are you raising now?" he demanded inquisitively, pointing to the twelve-foot concrete wall with the heavy, solid wooden gate, which had been erected all around the former chicken yard.

"Skunks," Brown briefly informed him.

"For their fur?" the pest persisted.

"No, for their perfume. The Chinese say it is good for rheumatism. I'm going to try it out on some of my patients."

And with that the doctor left the skeptical farmer scratching his head and returned to his breakfast. The housekeeper had been told a similar yarn, so that she should not feel tempted to feed the new pets in the doctor's absence. She was a kind-hearted soul, but prudence put limits to charity. Brown felt secure in his innocent deceit.

After breakfast he drove over to the university. Being on the program committee of the Biological Society, he wished to be at the auditorium well in advance of the meeting to round up the speakers.

"What's this stuff old Wilkes is springing this morning?" a somewhat flippant young man in rimless glasses demanded. "He's down on the program for a paper on 'New Light on Evolution.' Where did he get it? Wilkes hasn't had a new light on anything for the past twenty years. Do you think it will be worth hearing?"

" 'He that hath ears to hear, let him hear,' " Brown quoted with an enigmatic smile and passed on.

"Now what did he mean by that?" the cocksure young man muttered, unconsciously scratching his left ear. Wilkes' paper had been accorded the place of honor on the program, also the unusual time allowance of forty-five minutes. This was solely due to Brown's earnest persuasions with the committee. The other members took the doctor's word for it that he had carefully gone over the paper and that it was of the first importance. As ten o'clock drew near Brown began to show traces of nervousness. What if Wilkes overdid things and made it too sensational? The society would jump all over him. To ease his feelings, he fussed about the lantern and the motion-picture machine, heckling the operator with unnecessary directions.

"Run it through in slow motion first," the doctor emphasized for about the twentieth time. "Then give it to them as fast as you can without blurring."

"Sure, I understand," the operator replied gruffly. "You told me before."

"If you foozle it—" The doctor hurried for the platform. It was time to open the meeting, and the chairman was still smoking in the lobby. Just as Brown reached the door De Soto sauntered in and took a front seat.

"Hullo," Brown exclaimed under his breath. "He suspects someone. I hope it isn't me." Going over to De Soto, he gave him a hurried greeting. "Mixing a little biology with your physics?"

"Only a little," De Soto smiled. "I saw the title of Wilkes' address in the paper and thought it might be amusing."

"It will. The old chap has something brand new." He lowered his voice. "How is Mrs. De Soto this morning?"

De Soto's face clouded. "Nervous again. I made her take some of what you prescribed. Could you drop 'round to see her some time today?"

"Certainly. I'll go as soon as Wilkes has read his paper. Excuse me now; I've got to start things going."

The chairman regretfully flung away his half-smoked cigar, mounted the platform and called the meeting to order. As the business meeting had already been attended to by the council, he proceeded at once with the scientific program.

"The first paper is entitled 'New Light on Evolution,' by Professor Wilkes." He turned and nodded to Wilkes, who sat in the front row not far from De Soto. "Professor Wilkes."

Wilkes gravely mounted the platform. The curious audience of some three hundred expert biologists and intelligent amateurs with an interest in evolution noted that the professor had no manuscript in his hand. The experts sat back with a sigh of disappointment. After all it was to be a popular address, ninety-nine per cent. hot air and inspiration, one per cent. scientific fact. It was just what they would have expected from Wilkes, who had been scientifically dead for years. Wilkes gave them a surprise.

"Mr. Chairman, ladies and gentlemen," he began and proceeded at once to scientific business. "The first slide, please."

The hall was darkened and a beautifully executed microphotograph was projected on the screen.

"Old stuff," one skeptical expert whispered to his neighbor. "He got that out of Blair on the protozoa."

"Next," Wilkes requested.

"Blair's again," the skeptic whispered.

"Next."

"If he's going to show us all the protozoa in Blair, he won't get through till this time next year."

"Next."

"Hullo! Where did he get that one?"

"Next, please."

"Another, by Jove! Caught in the very act of dividing."

"Next."

"Fake," more than one expert whispered.

"Run them through more rapidly, please," Wilkes directed the operator, "one at a time."

As fast as he could change the slides the man at the lantern flashed on approximately a hundred photographs of the simplest animals known to science, the living things which consist of but a single cell. The exhibition was received in uncanny silence. Experts held their breath, amazed at the magnitude of what they were seeing—provided it was all genuine—or grimly waiting their chance to pounce on the audacious Wilkes should he prove to be hoaxing them. Laymen who had strayed into the meeting in the hope of witnessing a battle royal between monkeys and men felt vaguely disappointed. Why didn't the professor say something? The answer was simple. Words were superfluous to those who could read the pictures, and he was talking only to them.

The long series of individual slides came to an end.

"The motion pictures now, please," Wilkes requested. "Slow motion first."

The fruit of laborious weeks of toil by Wilkes and Brown was now slowly unrolled in a coherent sequence on the screen. The spectators saw a different succession of protozoa gradually evolving before their eyes. Types of the utmost simplicity survived through their transient generations, passed out of recognition as individual species and bloomed into new life, more complex and more highly specialized than their ancestors, and those again gave place to higher forms. The history of a million years flashed by every five seconds, and still the general trend was upward toward diversified perfection and increased richness of life. Gradually the rate of ascent slackened. The millions of years represented by sixty seconds of the moving film revealed no discernible variation in the structure of the minute, perfected creatures; they seemed to have passed forever into their perennial Golden Age. Then, in five seconds, first one splendidly developed organ degenerated, atrophied, and passed out of living history, then another, until within thirty seconds the descent was accomplished, and the countless millions of years of the slow, upward climb were undone. The whole cycle of evolution had swept round its circle, and the last generation, the end product of it all, was a degraded thing fit only to fasten as an inert parasite on the first creatures that had risen. Wilkes added a footnote.

"About one-third of the pictures from which that film was made are photographs, the rest are sketches by myself, Professor Hayashi of Tokyo, and a third man who wishes to remain anonymous. Please run it through fast now."

"One-third fake, two-thirds humbug," the skeptics whispered. "Wait till he gets through. He's the Charley Chaplin of biology."

The fast motion pictures were even more impressive. A whole race of animals seemed suddenly to open out like a rose in the sunshine, bloom gloriously in perfection for a few seconds and fade in a flash. The struggle of millions upon millions of years justified itself in those few seconds of beauty; the complete and final futility of the end mocked the struggle and made its justification a bitter nothing.

"Lights, please," Wilkes requested. "Thank you." He bowed, left the platform and resumed his seat. He had used but thirty-five of his allotted forty-five minutes.

"Is there any discussion?" the chairman asked.

A dozen men were on their feet instantly, but De Soto was first.

"Mr. De Soto," the chairman nodded. The others sat down on the edges of their seats.

"Mr. Chairman," De Soto began, "I must apologize for speaking in a biological meeting. But I should like to ask Professor Wilkes whether he has prepared a similar motion picture of the evolution of man."

"No," Wilkes replied, rising. "The data are not available."

"Do you think they could be obtained?"

"It is not impossible," Wilkes admitted quietly.

This was the last straw to the outraged experts. The chairman was forced to use his gavel.

"Professor Barnes," he announced when the commotion subsided.

Barnes was an unimaginative, middle-aged man who had made a very considerable reputation by contradicting his superiors on details of no importance and proving them in error on things which they had never said. If any disagreeable job was to be done, Barnes was the man to do it. The experts leaned back, satisfied that their case was now in competent hands.

"I fail to see," Barnes began in an injured tone, "why a meeting of the Biological Society should be turned into a vaudeville for the entertainment of amateurs. No competent biologist would give Professor Wilkes' fantastic reconstruction of what he imagines to be the past and future history of the evolution of the protozoa a moment's consideration. I move that Professor Wilkes be requested to withdraw his paper."

"Second the motion!" came from a dozen scattered points, like the cracking of snipers from an ambush.

"It has been moved and seconded that Professor Wilkes be requested to withdraw his paper. Is there any discussion? Professor Wilkes?"

"I have nothing more to say at present."

"Any further discussion?"

"Before we vote," the chief skeptic, Barnes, volunteered, "I should like to know what Professor Wilkes meant by his last remark. It sounded to me like a threat."

"Professor Wilkes?"

"Mr. Chairman, it is only fair to answer the gentleman's question. The society holds its next meeting three months hence in San Francisco. Six months from now we meet here again. At that time, if the gentleman still wishes further evidence, I will present him with an argument that would silence Balaam's ass."

All but two of the audience laughed. Brown noted that one of the exceptions was De Soto. The motion was carried.

"Professor Wilkes is requested to withdraw his paper," the chairman announced. "Professor Wilkes."

Wilkes rose up, lean, angular, self-possessed and obstinate.

"I'll be damned if I do," he said simply and sat down.

In the ensuing debate it developed that parliamentary law had no statute adequate to deal with Wilkes' offense and his unrepentant contumacy. To the chagrin of the conservatives, the paper had to stand as delivered. It would be duly printed, slides and all, in the society's sober proceedings. Several left the auditorium. Wilkes stuck to his seat like a barnacle, saying nothing, his jaw set and the light of battle in his eyes. He was not without sympathizers. Already the astute reporters were composing their sensational stories in which the name "Balaam Barnes" figured with undue frequency.

Brown nodded to De Soto and they left the meeting together.

"A great paper," De Soto remarked as they stepped into the doctor's car. "Where did Wilkes get his material?"

"Do you really want to know?" Brown asked seriously.

"I may as well," De Soto replied.

"Wilkes and I got most of it from the water in which you bathed that evening you came to my house."

To the doctor's surprise, De Soto showed no astonishment.

"I suspected it," he admittedly finally. "Do you and Crane know what you are doing?"

"Not exactly. Nor," he added after a pause, "do you know what you are doing. Otherwise how do you explain your obvious ignorance when you came to consult me? You did not know the cause of the itching of your skin."

"I admit it. But you forget that nature is like an open book to me—when I am feeling well. Something is happening to my brain," he continued after a long pause. "Thinking tires me. I never used to think, but saw the inevitable consequences of any pattern of circumstances—no matter how complicated—immediately, like a photograph of the future. The tingling of my skin did not puzzle me. I knew the cause."

"Then why did you consult me? You were pretty badly upset."

"I was and I still am. Through a stupid blunder I exposed myself to rays of an unknown hardness. If only two million volts or less, say, than five million, were being fed into my tube when I forgot to open the switches, I should not worry. In that case only the lowest forms of life would be affected."

"The protozoa on your skin?"

"Exactly. Have you any idea where those initial forms originate?"

"That is the one thing that puzzles me," Brown admitted. "The average human skin is alive with bacteria and other low forms of life, but not with the sort of thing we have just seen."

"Not even with lower forms that might evolve into the types in Wilkes' picture?"

"No. I am certain of it. The whole trend is different. Do you know what I suspect?"

"I can guess. You think that you and Crane have stumbled upon the secret of creating life. You haven't."

"Then what have we done? Where do those types originate?"

"In star dust. They are not those that have survived in the course of terrestrial evolution. But let that pass. What can theories do for me, with Alice on my hands?"

"You suspect something?"

"I do. And so do you. You have guessed by now, of course, that I know who Crane's confederate was in my laboratory that night?"

The doctor went cold.

"Who?" he asked in a level voice.

"You, of course. Couldn't you guess that human motives and commonplace human deceit would be childish games to a man who reads all nature as you read your newspaper? Or rather," he added in a low voice, "a man who once had that capacity."

"You are losing it?" Brown demanded quietly. "I thought I had noticed a dulling of your faculties. Why don't you rest? Your color is not good."

"I can't. But let that go; it is of no importance. To go back to the other for a moment. You men are all so trivial, so unambitious for anything that will count a million years from now. Laugh if you like. What good are the futile things you do for yourselves and your children? Think of the race—the human race! As individuals we are like those parasites on my body that Wilkes and you have taken all this labor to elaborate. The race is on my body; men, the protozoa swarming over it and breeding aimlessly. If we cannot preserve and mature the whole race and make one intelligent, purposeful being out of it, we are no better than an irritating itch on the skin of eternity. I could have done so much for it—once. They asked me for trash that would delight an idiot child for half a minute. They still ask it and I shall give it to them—till I get tired."

"Let me repeat," Brown persisted, "that you are ill and must rest."

"Don't I know it? Then why can't I rest? Just because I *am* unwell. When I first thought of marrying her, Alice was no more to me than you are or even Crane. She was just another human being. Some day I may tell you why I married her. Then a stupid accident began my degeneration. In another six months I shall be as foolishly humane as you are—curing the sick and helping the defective who should be mercifully exterminated or at least sterilized. I have grown to love my wife, even as you might yours, if you were married. That is why I have let you believe that you and Crane had deceived me. When you know everything, you will see that I am degenerated and done. Four months ago I could have solved my own problem. Now I can't. I have to rely on you."

"In what way?"

"Need she go through with what is before her?"

"I am afraid she must, even if I could throw my professional ethics overboard—which I can't. She is too far along. Why can't you speak out? Has the worst happened?"

"I don't know. I have degenerated, I tell you!"

"Well," the doctor muttered, "here we are. It can't be undone now. I'll go up and see her. She eats well and sleeps normally?"

"Yes, but she is afraid. Even I can see that she is not natural."

"It is only your morbid fancy, man. Cheer up. She will come out of this with flying colors, and you'll be the happiest man on earth."

The doctor found Alice happy and cheerful. The usual sickness had left her and she was busily fussing with the plants in the conservatory off her bedroom.

"Tell Miguel I'm all right now," she begged. "I can't bear being marooned here all night while he is off working at the laboratory."

"But I thought he was taking a layoff, Alice?"

"Oh, I know he is supposed to be having a vacation. But he spends practically the whole night at the laboratory and often most of the morning. Then, when he comes home all tired out in the afternoon, he is so cross I hardly dare speak to him." She smiled ruefully. "He seems to prefer the company of his pets to mine."

"You just imagine it. Don't you know that a woman in your condition always sees thousands of things that aren't so? Why, I was just talking with Miguel about you when we drove up. He's positively silly about you."

"Do you think he would let me watch him working in the laboratory again as he used to do?" she asked, brightening. "He loved it."

"Why not ask him?" the doctor suggested, eyeing her narrowly.

"Oh, he always puts me off. I sometimes wonder whether there isn't another—" She stopped, embarrassed by the accusation she could not frame.

"Woman?" Brown finished for her, laughing. "My dear Alice, you are like all the rest at this time. Tell your husband what you fear, and he won't let you out of his sight."

That evening at dinner De Soto, acting on a hint from

Brown, went out of his way to keep Alice amused and interested. She had a natural taste for science and was fairly well informed on all that went on at the Foundation. Biology, however, was an unexplored romance to her, as it is to most young women who should know it—if they should know any science. At school and college she had been fed the traditional slops of literature, economics, art and domestic science, with not one significant word of the one body of knowledge which women, above all others, should know. The vital functions of her own being were *terra incognita* to her, and the simple facts of the great miracle now transforming her whole life were as unknown to her as they might have been—and were—to an educated woman of the middle ages.

Her husband sought to enlighten her. He began with an amusing account of Wilkes' paper and its reception by the hopeless conservatives. Thence he launched out on a flaming prophecy of what mankind might do, were it so minded, with its own destiny.

"But," Alice objected, "fate or destiny is something that cannot be altered."

"In the past, yes. We have blindly let nature lead us. A century from now, if we wished, we might be leading nature."

"Is that what you are working on?"

"I was," he admitted in a strange voice. He rubbed the back of his hand across his eyes. "But I am forgetting how."

"Perhaps if I were to be your mascot again your luck would return," she suggested gaily. "Can't I come and watch you tonight?"

De Soto started to raise some pertinent objections, saw the hurt look in her eyes and yielded.

"Come along," he said cordially. "On one condition, however. You must take a nap whenever you feel sleepy. I shall probably be working all night. There is a comfortable cot in the closet there. I'll drag it out when you begin to nod."

She was absurdly happy. "Miguel," she confessed, "do you know what I was imagining in my morbid condition? Other men run about with women, and I feared you might get that way, too."

He laughed boisterously. "Women? I haven't thought of another woman since—" The puzzled frown that was becoming habitual with him suddenly darkened his laughing face. "Since when?" he muttered, scarcely aware of her presence. "I seem to remember a dirty suitcase full of letters. From girls. Where was it?"

"Buenos Aires?" she suggested softly.

"How could it be? I was never in South America."

"Oh, Miguel! Can't you remember anything? Where did you study? You knew all about physics when you first came to the Foundation. At least that is what my father told me."

"Did he? Then he must be right. I have forgotten. Never mind now. It really doesn't matter." He paused irresolutely before putting the question he vaguely feared to ask. "Did your father ever speak of anyone by the name of Wilson?"

"Not that I remember. Why do you ask?"

"Just a fancy. I seem to recall a man of that name who had a great deal to do with my education. Have you ever heard of amnesia—loss of memory? Well, I often think that is what is the matter with me. Some day my whole past life will come back. Honestly, Alice, it is all as black to me as it is to you. That is why I work incessantly—so that I shall never remember."

"You are afraid of what you have forgotten?" she asked quietly. She spoke to him as one might to an ailing child. He, not she, was the one in need of care.

"Desperately," he admitted. "Work is the only relief. Sometimes, do you know," he continued gravely, "I am so disturbed that I am tempted to try drink. Yet I have never touched the stuff."

"Don't," she counseled. "That would only make it worse.

Can't you remember anything of your father and mother?"

"It is all so impossible," he replied with a short laugh. "Did you ever hear how Leonardo da Vinci is said to have remembered, when he was a grown man, the days when his mother nursed him? You have read that? I go farther back in my memories. I remember the dark place where I lived before I was born. There was an intolerable flash of light, a terrible conflict in the darkness, and I found myself in a world that seemed strangely familiar yet utterly new. My very life contradicted itself; I had no right to live, and yet lived. Gradually the dark place of my prenatal memory faded and I found myself a man. The same thing is happening again. It is just like those slides that Wilkes showed this morning and his great moving pictures. Those protozoa I told you of slowly climbed to the very peak of their perfection, only to shoot to ruin in what, comparatively, was a second. The flower of my manhood has closed. Old age is upon me and beyond it the darkness of oblivion."

"Oh, Miguel! Can't you see that you are still a young man? Isn't your mind as fresh as it ever was—since you began your true life work?"

"My true life work? I have forgotten what it is. Only an aimless conflict of cross purposes remains. No sooner is a project started than I tire of it. There was one thing—or were there two?—that I hoped to do for the whole race. Did I ever start them? If so, I have forgotten. For all I know, both of them may now be working out for the good or evil of us all."

"Don't you have some definite aim in the work you are doing now?"

"Apparently not. I work by instinct and by habit to drug my mind. Without incessant work I should be forced to deaden my brain with drugs or drink. The most terrible part of it all is that one dead purpose after another speaks unexpectedly from its grave when I am alone and thinking of nothing. Then I try to put it into action, only to lose interest before I have completed a definite piece of work."

"Can't you find some one thing that will interest you and make you happy for its own sake?"

"There is one," he said slowly, fixing her with his sombre eyes. "And it is at the root of all the others, if only I could remember why it is."

"What is it?" she whispered. The look on his face made her feel old and ill.

"Life and what it may become," he answered. "The creation of life and the remaking of it to my will, in spite of chance and blundering evolution. This was my dream." He absently reached for her cigarettes, took one, put it between his lips, lit it and inhaled deeply like an inveterate smoker. "I have not yet told you the worst that rides me like a nightmare and makes me afraid to lie down at night." She was staring at him, round-eyed. "The worst is this. I know that I shall return to the black place where I lived before I had a mind, and I know that I shall remember everything when it is too late. One hideous thing that I cannot explain always comes out of the darkness when I close my eyes."

"What?" she asked, cold with fear.

"A black spider. This is the key to my lost memory."

She tried to hide the terrible shock his confession had given her.

"Miguel," she said, "I never knew you smoked."

"Am I smoking?" he exclaimed, staring at the cigarette as if it were a deadly viper. "When did I light this?"

"A minute or two ago. Don't you remember?"

"No! I have never smoked." He wiped his mouth distastefully. "The smell of tobacco nauseates me. That settles it. Time I was at work, instead of sitting here talking nonsense."

"I'm coming," she insisted firmly. "You said I might."

"Did I? Well, come along. Brown was right," he laughed.

"He told me not to let you out of my sight or you'll be getting foolish notions into your head. We'll have a good time; I'll work while you read and sleep. Come on; there's nothing like work."

They reached the laboratory shortly after nine o'clock. It was quite like old times. Work, after all, Alice thought with secret joy, was the one solvent for her husband's moodiness. Like many who tax their minds incessantly, Miguel was inclined to be neurotic. Creation was the only relief for him—self-forgetfulness. "Who would find his life must lose it." So she thought, poor girl, little dreaming that the self-torturing man at the dinner table was her true husband, and the brilliant inventor absorbed in his work the artificial shadow.

"I'm working at the two million volt level tonight," he informed her, "so I must be careful."

"But I thought you handled twenty million volts without worrying much," she objected with mild surprise.

"I do. The two million volt is the critical point. The slightest slip, and I pass from the gamma rays to the cosmic—the softer, of course. Once they start generating in the tube, they may go on indefinitely and rip through the whole scale, beyond the very hardest rays that come to us from interstellar space. Then there is likely to be the devil to pay unless we are adequately protected. So I shall make you wear a triple outfit of the screening material. It won't interfere with your movements. The stuff is as light as a cobweb."

"Are you sure you won't be in danger?"

"Positive," he laughed. "I've been working at this for two days now."

Going to the closet, he clothed himself in three suits of insulation and selected the same for Alice. Before taking the garments to her, he glanced furtively toward the chair where she sat reading, noticed that she was apparently absorbed in her book, and softly closed the door of the closet. Then, from a shelf beneath the electric light, he picked up a small flat dish the size of a silver dollar, and held it up to the light. The dish was full of water. In the water a single transparent globule, as big as a small pea, just floated. The globule might have been an oil drop or a fish egg for all that the uninstructed observer could see without close inspection. A slightly darker nucleus, however, precluded the oil drop hypothesis. De Soto seemed satisfied. He deposited the tiny dish on the floor, picked up the three suits for Alice, turned off the light, and opened the closet door.

"Here are your togs," he called, carefully closing the door of the closet. "Come over here and I'll help you on with them."

They were as happy as a pair of children. All the gloomy talk of the dinner table was forgotten in the simple adventure of dressing Alice up to look like a strayed aviator from another planet.

At last she was dressed in her triple armor and went back to her station. De Soto walked over to the black devil box and began making the connections with the new tube. This was not a replica of the one which Crane had smashed, but an improved design. It was indeed the very model which he had exhibited the previous afternoon to the desperate trustees as his answer—when fully developed—to the bid of Williams and Company for the power markets of the world. This, he had emphasized, was merely the key idea; the commercial development of it would be a work of months for the whole staff. But it would be ready when Williams shot his bolt. Experiments for the good of commerce and the salvation of the trustees, however, were not De Soto's object for the moment. His purpose was more abstract. One cannot always be thinking of money, especially when one has more than is necessary. The evening's work was to be devoted to pure curiosity.

It started tamely enough. The easy connections were made

almost automatically, and De Soto threw in the first two hundred and fifty thousand volts. Unlike Crane's unwieldy tube, De Soto's kicked up no spectacular display. There was no fluorescence. Alice followed his movements surreptitiously, saw that he was absorbed and happy in his work, and dipped into her book. By carefully timed steps he worked the voltage up to the two million mark and stopped. Alice glanced up.

"I must say your experiment isn't very exciting," she called across the laboratory.

He had completely forgotten her presence. At the moment she spoke, his back was toward her. Hearing a voice, he started violently. Then he remembered, and laughed. His wife was there. But in wheeling round he brushed against two of the screw switches with the sleeve of his transparent armor. The tube was set to receive and withstand only two million volts. Instantly, an unpredictable mishap, twenty million surged against the cathode with an irresistible impact.

It was too late to rectify the error by "killing" the whole apparatus. De Soto did this automatically when he realized what had happened, as he did immediately. Alice saw his face freeze in horror, why, she could not understand.

"Is anything wrong?" she cried, starting up and running toward him.

"Stay there!" he shouted. "Don't touch your clothes!"

"Come away!" she cried, dreading she knew not what. "Oh why don't you come?"

"I can't," he groaned, frozen where he stood. "I begin to remember. Watch!"

In the lower half of the tube a blinding blue light suddenly flashed up, flooding the laboratory with a ghastly, lurid brilliance.

"It should be white!" he croaked. "This is wrong!"

The blue light contracted, as if compressed by an invisible piston, and increased intolerably in intensity. Narrowing rapidly to a mere plane of blue fire as the piston descended, it became extinct.

"Look out!" he shouted. "It is going to explode!"

The concussion never came. Staring at the sheer black of the vacuum, De Soto saw the tiny vortices which he anticipated like a man in a dream, spinning from the outside of the crystal window and expanding as they spun. One broke against his protected hand, another struck the transparent insulation before his lips, and still he could not remember.

"I have done this before," he groaned. "Where? When?"

"Come away!" Alice entreated, seizing his arm. "This must be dangerous—oh! what is that?"

His eyes followed hers to the door of the closet. Something was moving about angrily in the darkness and blundering against the loosely fastened door in its efforts to escape.

"What is it?" she choked, clutching his arm in terror.

"I can't remember. There was a spider in a box—"

He never finished the sentence. The flimsy catch suddenly gave way, and the incredible monster lurched into the laboratory. Believing she had gone mad, Alice fled shrieking for the exit. De Soto froze where he stood, fascinated by the enormous creature hopping toward him. It was a toad, the size of a full grown man, hideously deformed, without eyes, its gelatinous skin pitted and pocked with holes the size of a human fist from which dripped and trickled a constant shower of young. As they rolled helplessly over the concrete floor the lumps of spawn began to develop, to thrust out feeble legs, and to increase in bulk like the arithmetic of a nightmare. The huge misshapen brute collapsed and became a swarming lump of fecundity.

Before he realized what he held in his hands, De Soto found himself playing the withering flame of the oxy-acetylene torch over the hissing mass and its multiplying offspring. As they puffed up and burst under the fierce heat, to disappear in wisps of vapor, he had a vision of thousands of black spiders boiling from a small box. It had happened once, but where?

Sick with loathing when his task ended, he rushed from the laboratory to overtake his wife. She had collapsed outside the door.

"I must make her believe it never happened," he groaned, lifting her in his arms. "Taxi," he shouted, hailing a passing driver. "My wife is unwell. Hurry!" He gave the address and tumbled in with her. "I am a fool," he muttered. "Like all of them I can only blunder."

On reaching home he put her to bed and telephoned for Brown.

"She had a fright in the laboratory," he explained. "Tell her it was nothing."

"Was it nothing?"

"Yes, if she is to keep her mind."

Alice lay critically ill for two weeks. During her waking moments she was barely rational. Whether she believed the assurance of De Soto and the doctor that she had imagined the horror, neither ever learned. When at last she recovered, pale and shaky, she never referred to the incidents of that terrible evening. They thought she had forgotten.

12

HIS SON

ONE MORNING SIX MONTHS LATER, A PUZZLED oculist sat staring into the right eye of a tired-looking young man of sallow complexion.

"It is the most extraordinary thing I ever saw," the oculist exclaimed. "You say your vision is still perfect?"

"As good as it ever was," the oldish-looking young man responded wearily.

"When did you first notice this?"

"About five months ago, one morning while I was shaving. I saw a small blue speck on the top rim of the iris—at the base of the blue wedge now. At first I thought it might be the beginning of a cataract. As I never read now, I didn't worry much."

"It is not a cataract," the oculist asserted. "There is simply a thin blue wedge in the general black of the iris. Your eyes are changing color, that's all. Nothing to be alarmed about. Would you mind my reporting the case to the Medical Society? Of course I shall not give your name, Mr. De Soto."

"Not at all. So there is nothing to worry about?"

"Nothing that I can see. You are just reversing the usual order. Babies born with blue eyes often turn brown-eyed or black-eyed after a few months. I hope you're not going back to the nursing bottle," he concluded with a laugh.

"No fear," De Soto responded gloomily. "But I should like a good jolt of whiskey."

"Perhaps I can oblige you," the oculist smiled. "I keep this for my patients when they must hear bad news." He poured a stiff drink for De Soto and half a dozen drops for himself. "Here's luck."

"Luck," De Soto responded, and tossed the drink down his throat. "That was what I needed."

Outside in the cool morning sunshine, he had a sudden revulsion of distaste. "What ever made me drink that rotten stuff? It tasted like varnish. Ugh! Never again."

He hailed a cab and drove to the Foundation. There was to be a full meeting of the Board of Trustees to discuss the offensive of the Erickson against the Universal Power Transmission Company. Disregarding Crane's hundred thousand dollar tip, Williams had gone ahead at top speed for the past six months developing "his" invention on a world-wide commercial basis. For the past month he and his associates had been deluging America, Europe, Africa and Asia with their propaganda, broadcasting the glad tidings that the wireless transmission of electrical energy—high power or low power—was no longer a dream of the theoretical engineers, but an accomplished fact that would shortly be on the market. This, as they justly claimed, was an industrial advance comparable in importance with the invention of the steam engine. Just as the steam engine with its railways and steamships killed the stage-lines and the windjammers at one swipe and brought about the industrial revolution, so this new method, as simple as A, B, C, of transmitting electricity without wires from producer to consumer, would stand the industrial world once more on its head and shake the last nickel out of its pockets. The nickels, the dimes and the dollars were already beginning to rain down in a jingling shower that threatened to drown the new company in a deluge of prosperity. Many a solid concern rated in the billion or half billion class had already thrown up its hands. Why fight? Their flanks were turned and their retreat cut off. Better to make peace while they might by selling out to the junk dealers and passing on the loss to their stockholders and bondholders.

Through all this furious publicity the Erickson crowd remained strangely silent. Was it the silence of defeat or the prelude to a stealthy, wholesale throat-cutting that wouldn't leave the foolish Universal a larynx to crow with? There was no doubt that Universal actually could transmit electricity without wires at a negligible cost and that it was prepared to do so throughout the civilized world. Then why was the Erickson so quiet about it all? Surely it must be ruined with the rest? Only the trustees and the technical staff knew the answer. Patents had been applied for but not yet granted. And the applications were so ingeniously framed that not one expert in a million would guess their particular value. Even the technical staff as a whole did not fully grasp what they were doing; De Soto so apportioned the details that no one man could possibly get a glimpse of the whole. The preliminaries were ended; he himself would put the finishing touches.

The president opened the meeting with a glowing tribute to the genius of their Director, who, he declared, had given them the world to play with. They must not, he concluded in tones of lofty solemnity, abuse the great privilege which their own business enterprise and the great skill of their Director had given them. Far from it. Greed and unscrupulous monopoly might actuate their competitors—witness the ruthless manner in which the Universal was crowding less lucky corporations into the ditch—but such base motives never had been those of the Erickson and never would be. A world monopoly not only of power transmission but of the means of generating power would put the Erickson beyond competition. Be theirs the mission to bring industry and the public—the ultimate consumer—into closer harmony and a deeper appreciation of the inestimable benefits which a wise business foresight confers upon groping humanity. All this for a reasonable and legitimate profit of a thousand per cent. on their investment. Would Mr. De Soto care to make a few remarks?

Mr. De Soto would. He swayed slightly as he rose to reply —for to him it was a reply, and not a mere footnote.

"Gentlemen," he began, "you must pardon me for being just a little drunk."

"Mr. De Soto!" the president soothed in an audible undertone. "We know you are joking."

"I am not joking. I'm drunk. Fifteen minutes ago I had a damned good stiff jolt of real whiskey. Otherwise I shouldn't be talking now. I had intended saying it sometime later. Please don't make the mistake of thinking I'm so drunk that I can't see straight. It was the first drink I've taken since—God knows when, I don't. And it has gone to my head. It gives me a warm, human glow."

They stared at him in astonishment. Was this their usually polite—if sometimes brusque—severely scientific and eminently practical Director? The man who had given them the world to play with? Surely not. And yet—*could* he be telling the truth? But then, he never drank. At their hospitable homes he had always waved the cocktails and the gin, the sherry and the wine aside with an air of sincere indifference that no amount of art could hope to stimulate.

"A warm, human glow, gentlemen," he repeated with emphasis. "Do you know what that means? You don't. Right now you are thinking of ways to quarter Universal after it is dead. You don't like its presumption in daring to invade your territory. Will you stop when you have broken them? You will not. As long as one of its backers, or its bondholders, or its stockholders has a dollar in his pocket, you are going after it till you get it. Pardon me, gentlemen, if I cannot restrain my feelings."

He turned aside and spat out of the window. Continuing, he made his plea.

"You make me sick. Sicker than that rotten whiskey made me. All my life I have been looking for a human being, and I haven't found one." His tone changed. "For humanity's sake," he said in a low voice, "I implore you to drop this before it is too late."

A trustee rose. De Soto's words had impressed him.

"Are we to understand that your answer to Universal is not what you thought it was?"

De Soto burst into a roar of laughter.

"Incorrigible," he shouted. "Absolutely incorrigible. Take what is coming to you."

The president took up the parable. Numerous disquieting hints released by Crane came home to roost. In particular he recalled Crane's disquieting theory that De Soto might ruin both the Universal and the Erickson, not merely one or the other.

"You feel confident that our demonstration will convince the experts?"

This time De Soto did not even smile. His plea, he realized, would fall on stones.

"It would convince anyone," he said. "Send out your invitations for four weeks from today—cable, telegraph, write. That will give the Europeans ample time to get here. Don't forget to include liberal travelling expenses and expert fees. You will get it all back. But, for the last time, I ask you to call off the whole thing. There is one humane thing to do now, and only one. Lay your whole project before the Universal. They will see that they are hopelessly beaten. Then agree to withdraw your scheme, scrap the invention, and forget it completely, if they will do the same with theirs. They can lose no more that way than if they stick to the last. If they ever attempt to market their device or to transmit power themselves you can stop them instantly by threatening to compete."

"But what is the point?" a trustee obejcted.

"I can't tell you."

"Why didn't you warn us—as you seem to be doing now—six months ago?"

"Because then I had not gone soft. My plans were different, although even then I was beginning to doubt and to weaken."

"Weaken on what, Mr. De Soto?" the president demanded curiously.

"My purpose when I first sought employment at this Foundation."

"And what was that?"

"I will not tell you."

"Why not, Mr. De Soto? It cannot have been dishonorable, surely?"

"Dishonorable?" De Soto laughed. "What is honor to a fool? I do not choose to tell you because I have changed my mind. Or rather," he added, "my mind has changed me."

They scrutinized him shrewdly. Was he trying to betray them to Universal. At length one trustee expressed the common sentiment.

"If, as Mr. De Soto assures us, we can't lose out, I don't see why we should discuss the matter further. I move that the Director be instructed to carry out his program, four weeks from today, as already arranged."

"Second the motion."

"Moved and seconded—."

The vote was carried unanimously.

"This is your final action?" De Soto asked quietly. The president nodded. "Then I shall make my last appeal. You will smash Universal, as I have promised. But in doing so you will not benefit your customers. Have you their interests in mind, or your own? I can convince Universal that it also will not make its customers any happier. If you abandon this now, I will make them give up theirs tomorrow. Which is it to be? Your own gain, or that of the people you serve?"

There was a dead silence.

"Very well," De Soto continued. "That is your answer. I understand. Please accept my resignation, to take effect immediately. After all, I shall have accomplished my initial purpose in joining your staff. Perhaps it is the best. I tried to nullify it only because I have gone soft. You yourselves are the best judges of what is best for you."

"Don't act in haste," the president begged as De Soto walked from the room. "We shan't accept your resignation until you have had four weeks to think things over.

"Take a rest and you'll feel better."

"Four weeks?" he echoed with a bitter smile, his hand on the door knob. "Why keep me? The full instructions for capturing the world markets in everything, not only power, which is at the bottom of it all—are already in your hands. Your Board has the detailed plan before you, and your very competent engineers can execute it. Put it into action four weeks from today. You will not need me. I shall move out of the Foundation residence tomorrow."

"Mr. De Soto!" the president protested in a shocked voice. "The residence is yours indefinitely, whether you stay with us or not. Surely you do not think us—"

"I think nothing whatever about you," De Soto retorted, opening the door, "except that you are hanging yourself, your sons and your daughters, and saving me the trouble. I would tell you to go to hell, if I did not know that the next thirty years on this earth are going to beat any hell ever imagined by the worst diseased imagination of the middle ages—Dante's." Closing the door behind him, he left the outraged trustees to their thoughts.

"Drunk?" one hazarded.

"Or crazy. It will be a good thing if he does resign. We don't need him any longer, with this in our hands. I vote we make no advances to him to reconsider. What can he do for us? Nothing."

And that seemed to be the general opinion. The meeting dissolved without formal action on the resignation of the Director. Watch and pray, wait and watch, are good slogans, in business as elsewhere. They decided to watch. For the moment they would take a firmer grip on the world's tail and flex their flabby muscles for the luscious twist.

On reaching home, De Soto at once told his wife of his resignation. Alice was pale and ill. She listlessly acquiesced.

"You know best," she said.

"I plan to move out of this house tomorrow. It would be impossible to continue living here practically on the Foundation's charity. Let us move out to the country—I'll find a nice place."

"Can't we wait till—?" She did not complete the sentence. "It won't be long now."

He glanced at her, something like fear struggling with pity in his eyes.

"Certainly. We can stay here at least four weeks, if you really wish it. What does Brown say?"

"He hasn't called today. Doubtless he is busy."

"Yes, I remember. The Biological Society is meeting today and he is on the committee. Don't you worry. Everything will come out in fine shape. We can stay here indefinitely if you like—it was your home for years. Perhaps we had better."

"But your resignation?"

"Oh, that. I can do what I please with the trustees. If the worst comes to the worst, I'll buy the place. We're almost indecently rich, you know," he laughed, trying to cheer her. "That's why I resigned."

All her sparkle was dead. "I wish it were over," she sighed.

"There, there! You'll soon be as happy as a queen."

"Tell me, Miguel," she said slowly, "has my mind been right since that evening in the laboratory? Sometimes I seem to be living in a horrible dream. I fainted, didn't I? Do I seem rational to you?"

"Why shouldn't you?" he asked with assumed astonishment. "You are. These fancies are natural to you at this time. They mean nothing. Ask Brown when he comes, if you think I'm just talking to disguise the truth. What I tell you is cold, scientific fact, and he will back me up."

"I wish he would come."

"If that's all you're worrying about, it's soon cured. He will be here in fifteen minutes if he's still alive."

He left her to telephone to Brown's office. The doctor was at the Biological Society but was quickly reached and promised to come at once.

"She's imagining things," De Soto informed him in the hall. "Cheer her up."

"I'll do my best. They often get like that at this stage. It means nothing."

Thirty minutes elapsed before Brown rejoined the anxious husband.

"Well?" De Soto inquired.

"She is normal, except in one thing. Her mind seems to be straying."

"In what respect?" De Soto paled beneath his fast-fading tan.

"Sit down. I want to tell you something that I have never had the courage to confide to another living man—except Crane. He and I saw it together. If Alice is losing her mind, her delusions have a peculiar quality of truth. At least that is how I feel. Perhaps you will agree, when you have heard what I have to say. Ready?"

"I'm ready. You have seen the effect of the hardest rays on living tissue?"

"Yes. Crane and I together." In five minutes the doctor gave De Soto a sufficient account of what he and Crane had witnessed in the twenty million volt laboratory. "Those spiders," he concluded, "had evolved, bred and multiplied at a terrific rate in less than twenty-four hours. What accelerated their rate of evolution beyond all reason? Millions of years were compressed into those twenty-four hours. Where did those swarms of voracious brutes obtain their food? These are some of the questions that Crane and I think you can answer."

"Why do you think I should know? I wasn't connected with the Foundation when this happened. Although," he added with a bitter smile, "the solution of your problem is no more difficult than Crane's. He could not see how an anonymous letter might be delivered so as to arouse no suspicions among a pack of dull-witted drudges—cooks and waiters of the business world. Yours, I admit is a less trivial problem. Suppose things happened as you say they did. Where did those spiders obtain their food to make possible their greatly accelerated rates of evolution and development? They sanded the floor, you say, with millions of unhatched eggs. The mothers, at least, must have been well nourished. Did it never strike you that the same short wave rays which started the surge through all evolution for your spiders could also provide them with the necessary food? The nitrogen of the air, the carbon dioxide, the oxygen and the traces of noble gases were instantly aggregated into complicated organic compounds, based on the electrons positive and negative, under the influence of those rays. If matter can be utterly annihilated, or as 'miraculously' created out of the wandering protons and electrons by the hardest cosmic rays, might not the softer induce chemical changes, making food from the air? It is done even in our stupid laboratories. But it may all be a dream—I don't know."

"You admit that it is not unreasonable or absurd?"

"You wouldn't after seeing Wilkes' demonstration on the protozoa! A different set of cells were effected in your spiders; that's the only distinction. The hardness of the rays—or, if you prefer, the shortness of the waves in the radiation emitted, determines what cells will be stimulated or destroyed. You have guessed that much?"

"More, as you may see tonight, if you care to come to the public lecture. Wilkes is to talk again."

"And silence Balaam's ass?" De Soto suggested with a sardonic smile. "It can't be done. I tried this morning to answer several, and left them still braying. Still, if Alice is well enough and won't miss me, I'll be there to see the fun. She suspects that what came out of the closet was real, and not the creation of a sudden nervous breakdown?"

"Suspects? Alice knows that it was real. And what is darkening her mind is your silence. Why did you do it?"

"A pure blunder. I'm always blundering. Alice spoke and startled me. My sleeve did the rest. What happened was as much of a discovery to me as it was to her. Since that evening I have studied the effect exhaustively. If you care to inspect my menagerie, you will see that the last cage is empty. I'm done—beaten. I'll never use the chloroform bottle or the oxy-acetylene torch again. Nature has got the better of me at last."

"But what on earth did you think you were trying to do when you blundered?"

"As you refused to help me," De Soto replied grimly, "I tried to help myself. If I could control evolution in one direction, why not in the opposite? Then I could undo what you, as well as I, believe may have happened."

"And you found you could not pass up or down the scale at will?"

"No longer. Ten months ago I could have played on it like a flute—and I did. Now I have lost my capacity. Can't you see that I am degenerating? Look at my right eye. Is that blue wedge a normal change in a healthy man?"

Brown peered into the affected eye.

"When did this begin?"

"Nearly five months ago—more or less. Can you explain it? No? Neither can I. Nor can I account for my washed-out feeling. Do you notice my color? And the deadness of my hair?"

"All that is merely lack of tone due to overwork and worry. As soon as Alice is safely through you will be as good as ever."

"Better," he said bitterly. "That night when she saw the thing, I told her that I was going back to the dark place where I was born. Your account of what you did with those spiders is like a hand pushing me into the darkness. Something will rush out of it presently and destroy me. But before it does, I shall see the light I have been groping after for months." He brooded in gloomy thought for some moments without speaking. "Promise me," he said, looking straight into the doctor's eyes with a flash of his old dominance, "that you will take care of Alice whatever happens to me. I have loved her, and that has been my ruination. All of my business affairs are in good order. She will be wealthy. See that she is not fleeced and keep her from marrying some scum who wants only her money."

"Look here, De Soto," the doctor retorted quietly, "you mustn't think of anything like that. In spite of all that she half suspects, you are still her one reason for living."

"But if I die—naturally?"

"You won't, for years yet. However, I'll face it. If anything happens to you, I will see that your wife gets a square deal. Now let us talk of something more cheerful. Coming with me tonight to hear Wilkes' paper? There has been nothing like it in the history of science. Not a soul but Crane, Wilkes and me, and possibly you, has any idea of what the old chap is going to spring on the skeptics. We have played at least one bar on your magic flute."

De Soto brightened. "I'll be there," he laughed. "But wait till you hear the full orchestra."

"When?" Brown demanded.

"Thirty years from now. It will begin four weeks from today."

"What do you mean? Have you—."

"Wait and see. I may be dreaming."

"Then Crane and I must awaken you," the doctor retorted. "Don't fail to come tonight."

"I won't. *Au revoir.*"

The auditorium for the evening lecture was crowded. De Soto found a seat in the rear. Brown presided, to introduce the speaker. News had leaked out through the committee that the address of the evening, "New Light on Evolution" was likely to prove exciting. The newspapers had spared neither conjecture nor innuendo to advertise the meeting. Some even hinted that the long-missing link was at last to be exhibited to shut up the Fundamentalists. Others recalled the incident of Balaam's ass and wondered whether he would be present in person.

At eight fifteen exactly, Brown briefly introduced the speaker. This time Wilkes had a manuscript in his hand. He began by dryly reviewing the theory of evolution. Sensation hunters yawned and shuffled their feet. Wilkes paid not the slightest attention, but continued to bore through his dry-as-dust argument like a beetle in a board. At last—after forty minutes—he had finished his preliminaries. Tossing the manuscript aside, he squared his shoulders, adjusted his cuffs, and let loose without notes.

"All that, ladies and gentlemen, is old stuff. You learned it in the grammar school—or if you didn't, you should have. Evolution is less a theory than a description. Does it assign any *physical cause* for the origin of species? It does not. The facts which it is alleged to coordinate are almost as complicated as the theory which strings them together. Compared to any of the greater mathematical or physical theories, it is rather a childish effort. It does not go to the root of the matter."

The papers reported the next morning that there was considerable disorder at this point of the professor's address. Undismayed by the boos and jeers of the scientific fundamentalists, Wilkes raised his voice and kept on, disregarding Chairman Brown's frantic appeal for order.

"Old stuff!" he shouted. "As old as Democritus and as dead as Lucretius. Metaphysics, ladies and gentlemen, metaphysics! Until we can control the course of evolution in our laboratories we are no better than Aristotle with his cock and bull."

"Can you control it?" a ribald voice from the back of the hall demanded.

"Order!" Brown snapped. "There will be an opportunity to ask questions after the lecture."

"Since the disorderly gentleman in the rear has asked a pertinent question, I will make an exception, and answer him. No. I can not control evolution."

"Then what do you think you are talking about?" an infinitely dismal, sepulchral croak from the gallery inquired.

"Listen, and you will find out. Another interruption and I leave the platform."

There was a dead silence. A man in the front row, showing a disposition to chatter, was promptly squelched by his wife. Wilkes continued his extempore discourse.

"Facts first, fun later. Before you will be in a fit state of mind to appreciate my clinching argument and enjoy the fun—such as it is—I must get some hard dry facts into your heads. It may hurt those unaccustomed to using their brains, but nobody will be seriously injured.

"First, there is the cause, the physical reason for evolution. What is it? I don't know, and neither do you. Like Newton, 'hypotheses non fingo'—I don't indulge in wild guesses. But, like all scientists, I guess as Newton did. Then I check up my guesses against the facts, or against the experiments predicted by the guess.

"In this instance," the professor continued with evident relish, ignoring the drowsy blonde at his right in the third row, "in this instance the ascertained facts of paleontology are indisputable to all but Fundamentalists. We human beings are mammals—the female suckles her young, and our young are born alive. Reptiles are not mammals. For one thing their young are born only half alive, as eggs. Nor are birds mammals. Yet birds and mammals both sprang from the reptiles. That is the incontrovertible record of paleontology.

"Circle squarers, believers that the earth is flat and that the moon has no rotation, swarm in our midst, as the late Professor Tait observed on an occasion similar to the present. Modern statisticians have found that one person in five thousand believes he can square the circle, one in five hundred that the earth is flat, and one in five that the moon has no rotation. Therefore, I conclude, what I am about to say will be distasteful to my present highly cultured and intelligent audience. After all, science *may* be wrong, and the moon *may* be made of green cheese.

"Any human being who cares to go back far enough will find his family tree to be a mere twig on the greatest tree of life this earth has ever known, that of the reptiles. In short, the reptiles were our ancestors."

A prolonged hiss from the third row broke the thread of the professor's discourse. Wilkes paused appreciatively until the objector ran himself out of steam.

"Ah," he resumed, "I perceive that evolution has still a long way to go for some of us. To continue what I was saying before the gentleman in the third row obliged me with a practical demonstration. Suppose we *could* control evolution, both backwards and forwards. Imagine first that we can reverse the natural progress of man, and that we can do it at a greatly accelerated pace. In half an hour we should see ourselves chattering in the trees with our cousin apes; an hour would find both us and our cousins on familiar terms with queer little mammals that none of us would recognize as our great, great grandfathers; and finally, after about two hours of this prodigiously fast sweep into the 'backward abysm of time,' you and I would behold a strange and pathetic sight. We should see a bewildered colony of reptiles, their short, feeble arms clutched about their narrow bosoms, contemplating in horror and awe their unnatural broods—the first mammals. Could these unhappy parents look far enough into their misty future, they would see the last of their kind being mercilessly exterminated by the lusty descendants of these first, puny mammals.

"Before turning to a brighter picture, let us glance at another, more flattering to our human conceit. Suppose a common hen, or any other bird, could be sent back along the path which it had taken from the beginning of time. It would reach the reptiles much faster than we. Almost in a quarter of an hour—at the same relative speed as our own trip to our family tree—the hen would perceive that its feathers had given place to scales, and its toothless bill to a vicious, horny mouth crammed with long, sharp teeth.

"Now for the brighter picture. Accelerate the rate of evolution forward. What becomes of us? Ultimately, of course, we shall probably become as extinct as the great reptiles from which all our kind originally sprang. But, on the way to extinction there is one not wholly unpleasant prospect. We shall subdue the physical forces of nature almost completely, and the entire race of mankind will become incomparably more intelligent than it now is, with a greatly heightened joy in living. The discontented will have perished. That they may be noble in their discontent does not concern us. They will have gone the way of the dodo long before the race begins to live, for the simple reason that discontent is a destroying influence. It is nature's anaesthetic to drug the misfits into a readier acceptance of the death which is their one answer to a world with which they are unfitted to struggle. Many of them may be remembered for great work, for a little time, but they themselves, and in the end their work also, will perish.

"That is at least a not improbable conjecture. A second possibility that the future holds for us is equally obvious. Just as the mammals sprang from the reptiles, so from the mammals in turn, man included, may spring a totally new race of creatures. It is even possible, from minute examination of the germ cells of our own bodies, to predict in its broadest outlines what the race of our successors may look like. I shall not bore you with these speculations now, as facts that can be seen, heard and handled are more convincing to those who have eyes to see, ears to hear, and fingers to touch.

"Six months ago, I showed before this Society a series of drawings and photographs from life, in which it was proved that my associates and I had succeeded in compressing the whole evolution—millions and millions of years—of certain species of the lowest type of animal into a few hours. Those protozoa, beginning with the humblest, passing to the highest, and again sinking to the very lowest through innumerable generations, all within the short span of less than twenty-four hours' actual experiment, should have convinced those capable of human reason that my claims are valid. Was anyone convinced? No. 'Though one rose from the dead,' they would not believe. Hence I have prepared a more convincing demonstration, this time in the opposite direction. Rather, my friend has prepared such a proof; I am merely the showman. He has a professional reputation to lose; I never had one worth considering.

"Ladies and gentlemen, I now present you with a proof that we have succeeded in reversing evolution. What I am about to show you illustrates our process as applied to birds. If we are able, as we claim, to reverse evolution for the birds, we should be able to produce the prehistoric reptiles from which the birds sprang.

"Our starting point was a common brown hen of the Buff Orpington variety. Until she was subjected to the proper influences, she laid excellent eggs, many of which were eaten and enjoyed by my collaborators. They were normal hens' eggs. After our experiment, she began laying very small eggs. They were not shelled, but encased in a porous membrane like tough skin. Eighteen of these abnormal eggs hatched. I now show the reptile which hatched out of one of those eighteen eggs. It was a few days over four months of age when it died. Mr. Chairman, if you will have the alcohol tank wheeled onto the stage we can proceed with the demonstrations."

As one the audience rose to its feet. Those in the rear stood on the seats; those in the very front were restrained by guards from climbing to the platform.

"Everyone will have an opportunity to see the reptile," Wilkes shouted. "Please do not come up to the platform until the guards permit you to file past."

An oblong box like a coffin, draped in gray tarpaulins, was now wheeled onto the stage beside the speaker's stand.

"A little to the right, so that the whole house can see without interference," Wilkes directed. "That's it."

With a pardonably dramatic gesture, the professor unveiled his masterpiece by flapping off the tarpaulins.

"There!"

Submerged in the glass tank of alcohol a long, lemon-yellow monstrosity, like a huge lizard with an over-developed head, lay supine on its spiny back. The enormous head rested with its flat occiput on the bottom of the tank, its long, gaping jaws almost projecting above the level of the alcohol. From tail tip to head the reptile measured between eight and nine feet; its evil jaws could have crushed a young pig at one snap. The teeth, in double rows on both upper and lower jaws, might easily have crunched to fragments the bones of a large dog. The hind legs, like a crocodile's, were muscular and well developed; the front, mere fins with claws, were clasped pathetically over the narrow chest in the eternal resignation of death. The skin could hardly be called scaly. Rather it was a compact weave of triangular warts, each about the size of half a postage stamp. About the rigid jaws of the dead reptile lingered the frozen remains of a sardonic smile, as if the creature had looked both before and after, and was now as wise as a god.

Seeing is said to be believing. Those who assert that it is do not know either the scientific mind or the fundamentalist. The pickled reptile was received first with the silence of incredulity. Then, in ludicrous unison, a rhythmic chant of "Fake! Fake! Fake!" shook the auditorium. The crowd filed up to the platform, hustled by the guards, passed before the glass tank, saw with their own eyes the yellow monstrosity in the alcohol and doubted. This is not set down in any critical spirit; it is merely recorded here—as it was in the late extras that night—to show that the average human being is not the sort of fool who believes that seeing is believing. The skepticism of the crowd did it enormous credit. Through all that unsympathetic hour, while the irreverent humans filed past, the prehistoric reptile smiled his enigmatic smile like a cynical Pharaoh lying in state. He knew all evolution now, both forward and backward; his belated descendants some day would be as omniscient as he.

At a gesture from the chairman, the crowd at last resumed their seats. Brown made a brief address.

"In conclusion, I may say," he remarked, "that Professor Wilkes is not surprised by your reception of his evidence. May I ask for a show of hands? Those who consider this thing in the tank as substantiating, in some slight degree, Professor Wilkes' contention that evolution may be reversed by man, will please raise the right hand."

Several hands—at least a dozen—shot up. Before the meagre count could be taken, an indignant voice claimed the privilege of the floor.

"Mr. Chairman!"

"Mr. Barnes?"

"I must protest against a meeting of the Biological Society being turned into a revivalist experience orgy. Professor Wilkes has tried to foist upon the lay public a gross imposture. His so-called reptile is dead. I deny that any such reptile ever lived. The majority of those professional biologists now present—whose spokesman I have the honor to be—pronounce the yellow thing in that tank to be an extremely able fraud. Whoever has spent weeks, possibly months, in manufacturing that fake, might well have employed his talents to better advantage. That it is an almost perfect restoration of an extinct reptile of the middle period of the great reptilian race, which flourished on the earth ages ago, we do not deny. We merely assert that it is a forgery."

When the applause, foot-stamping and shouting, which greeted this fearless indictment of the exhibit in question as a bold fraud, had subsided, a resonant voice was heard claiming the chair's attention.

"Mr. De Soto?" Brown invited.

The whole audience turned to stare at the world-famous inventor standing up at the back of the hall. Until now he had passed unnoticed, save by those in his near neighborhood. What would he say? Light into the audacious Wilkes as Barnes had done?

"May I ask how Professor Wilkes induced the change in the germ cells of his hen to obtain this result?"

"As Professor Wilkes' paper has been received unfavorably," Brown replied, after a consultation in an undertone with the professor, "he prefers not to state for the present."

"In that case, Mr. Chairman," De Soto replied, "may I have the floor for five minutes?"

Brown's decision was drowned in an uproar. "De Soto! De Soto!" the crowd chanted. Brown at last got order.

"Mr. De Soto will take the platform in a few minutes, if he will be so kind. In the meantime, Professor Wilkes wishes to add a few remarks. May I ask you to keep your seats while he is speaking? Professor Wilkes."

The professor began in his driest voice.

"Any good scientist enjoys being called a liar by his brother scientists. It puts him on his mettle. I have to thank Professor Barnes for having performed that service for me. You will recall that I said eighteen of the reptile eggs produced by that brown hen hatched. One of the reptiles died a few days over four months of age. From the beginning he was the puniest, and we despaired of keeping him alive for a week. In spite of all we could do for the poor creature, he died. I, myself, sat up with him anxious nights, trying to nurse him back to health. Had Professor Barnes lavished his own maternal care as I did mine on that unhappy child of the prehistoric past, he would not scoff at its pitiful inadequacy now. That reptile, ladies and gentlemen, was kindly and affectionate—provided you kept out of reach of his teeth. I grew to love him more and fear him more than I love or fear Professor Barnes.

" 'Eyes have they and see not; ears, and hear not.' Of such are Professor Barnes and all his followers, the fundamentalists of biology. They call my poor dead friend a fake. I wish they were right, for then I might, as an artist, rival nature herself."

"You haven't answered him," a hollow voice from the second row suggested.

"You are right," Wilkes admitted. "I cannot answer him. His kind is unanswerable. For the rest, however, I have a little surprise, as a reward for their patient faith. Remember, eighteen eggs were hatched. One of the reptiles died; you see him here. What of the remaining seventeen?"

"Yes!" Barnes shouted, leaping to his feet. "What of them?"

"Kindly address your remarks to the Chair," Brown suggested acidly. "Professor Wilkes, what of the remaining seventeen?"

"They are alive and well," Wilkes replied simply.

"Show them to us!"—This from the audience at large.

"Unfortunately I cannot do so," Wilkes admitted regretfully.

Shouts of "Fake!" all but drowned the disappointed "Why?" Ignoring the former, Wilkes satisfied the latter.

"Because this stage would not hold all seventeen of them, or even six. They are not pleasant to handle outside of a steel cage. So I have brought only one, which my friend, Doctor Crane, will now show you. Doctor Crane."

Oh's and ah's bathed the professor like incense. He had scored his point, and Balaam Barnes was about to be silenced—at least they hoped so. For the average audience, scientific or other, is about as fickle as a flame.

The gorgeous purple velvet curtains parted at the back of the stage, revealing Crane in the act of bossing eight brawny workmen who tugged and hauled at an enormous cage of steel bars mounted on two low trucks. The cage was wheeled into the center of the stage; Crane withdrew with his workmen; the purple curtains closed; Wilkes followed Brown from the platform, and the guards braced themselves for the onset.

At first there was silence. Then fear. Then astonishment. Then a foolish, fluttering applause that died instantly. Again silence, tense and heavy with fear. The sluggish reptile in the cage raised its enormous head, stared for an uncomprehending five seconds at the pink and gray sea before it, regurgitated, and unconcernedly turned away. As the horny lips snarled back, the breathless spectators saw two double rows of cruel teeth, sharper than a shark's and as long as a sabre-toothed tiger's, bared for the attack.

"About six and a half months old," Wilkes remarked dryly. "Has Professor Barnes any comments?" The deathly silence remained unbroken. "If not," Wilkes continued, "you will presently notice a characteristic odor. Those of you who have ever smelt a large living snake, say a boa constrictor, will recognize the odor in a general way. This, you will admit, is similar, but much more intense, with qualities of its own. You are smelling, ladies and gentlemen, the same smell that paralyzed our mammalian ancestors with fright when they tried to hide from their reptilian parents in the reeds. Familiar, isn't it?"

There was no reply. The indifferent brute with a brain no bigger, perhaps, than a baby sparrow's, raised its head and preened its scales. Along its spiny backbone, and over its massive flanks, a riffle of triangular flecks of bright green passed lightly, like the sudden rubbing of an armor of artichokes the wrong way by an invisible hand. The crowd shuddered. They had seen birds do that. With a sudden movement of its sinewy neck that was almost graceful, the squatting brute ruffled the upstanding scales under its armpits, rapidly combed its backbone with its chattering teeth, shook its whole body luxuriously, and settled down to indolent ease. A cold, foul odor wafted over the audience.

"I refer you," Wilkes remarked from his station by the stage steps, "to any competent treatise on paleontology for the original of this reptile. It is well known. At least we think it is. When full grown, it will be as tall as a giraffe and as bulky as an elephant. This is one of the later species of reptile. The great ones had already begun to fade from the screen of evolution when this one thrived. Notice the degeneration of the thorax. It has an obviously inadequate lung capacity. The hind legs also show weakness. On the flanks—not where one would naturally expect to find such things—you will see two serrated excrescences. These are the rudimentary wings trying to break through the tough armor of scales. The wings, of course, did not originate in this way. What you see is merely one of nature's innumerable hit and miss, blunder and succeed methods of evolving a new species. The wings that finally came in were of deeper origin—in the mutated chromosomes of our friend here. Her—this reptile is a female—her germ cells contained the irresistible mutation that finally gave us the birds, including the brown hen from which this terrific beast was born.

"I call her terrific," he continued, "because I personally am very much afraid of her. You observe that she is just like her dead brother in the glass tank here, except that she is five times his size in linear dimensions, and therefore one hundred and twenty-five times his bulk. Before I state why I am afraid of her, let me assure everyone in this audience that neither I nor any of my associates has ever fed her a living animal. The pigs and calves which so far have kept her alive were duly slaughtered and butchered before being introduced into the cage. A pet cat belonging to one of my collaborators chanced to stray within reach of the cage one day when this reptile was much smaller and could get her head through between the bars. The cat was nipped. It died in four seconds. Now, for all that paleontology can tell us, we do not know whether or not this prehistoric reptile was venomous. From what happened to the cat, I suspect that it was. We shall have to wait until one of the seventeen dies. For sentimental reasons I do not care to dissect my dead friend in the tank. I shall ask you not to attempt to inspect the reptile on the stage. She may be dangerous."

As if to underline the professor's remarks, the huge mass in the cage suddenly became a spitting fury. Hurling herself against the bars of the cage, she slavered and screamed at the audience in an access of reptilian fury. Women fainted and were carried out; men stood their ground and tried not to show the white feather under the intolerable, nauseating stench which filled the auditorium as the raging half-snake, half-bird, lashed herself with her muscular tail and clawed at the steel bars with five-fingered hands that were strangely human.

"Take her out!" Brown shouted from the floor. Crane rallied his crew. Three minutes later the hissing screams were suddenly cut off by the closing of steel doors behind the purple curtains. Brown remounted the platform.

"Mr. De Soto wishes to say something," he announced.

The crowd sat down in dead silence, and De Soto walked to the front of the hall. Refusing Brown's invitation to mount the steps, he began speaking in a low voice, which carried to the farthest corners of the galleries.

"I state facts," he began, by way of introduction. "Several months ago I began to experiment on living tissue with high-frequency, short-wavelength X-rays. The results were encouraging. I used what knowledge I had gained from these preliminary experiments to predict what must happen under the influence of cosmic rays—the rays of shortest wavelength known to science. These rays will penetrate forty feet of solid lead. With this penetration they should be capable of affecting the smallest cells in all animals—insects, mammals, protozoa, man. By properly modulating the wavelengths of the rays sprayed upon the chromosomes, I found it possible to accelerate normal evolution or to retard it; to produce mutations—the creation of new species, such as mammals from reptiles—or to inhibit them. Perhaps here I overstate; my completed experiments do not fully justify my last assertion. I undertook, many months ago, to put my theories to a crucial test. Unfortunately certain accidents, due entirely to my own carelessness, make the outcome doubtful. I can only await the decisive answer which, I anticipate, will be given within the month. You agree, Doctor Brown?"

"I think so," Brown assented in a voice that was scarcely audible. De Soto nodded and went on.

"It is possible, I assert, to control evolution in both the forward direction and the backward. Professor Wilkes' two exhibits—that of six months ago with the protozoa, and that of tonight with the hen-reptile—put this beyond dispute. You have seen it with your own eyes. Even Professor Barnes, legitimately skeptical, has no further objection to offer.

"All this, ladies and gentlemen, is purely academic. It is of interest only to professional biologists. Of what application can it possibly be to you?

"Let me tell you. If we can control evolution; if we can hasten nature forward at the rate of a million years in one of our human years; if we can perfect our race as Professor Wilkes has predicted, who will profit? Who? Is it worth perfecting? I confess that I do not know.

"Suppose you were given the chance to perfect yourselves. Would you take it? I think not; for no one of us knows what perfection is.

"Suppose again that you were offered the opportunity of settling all of your problems, once and for all, within one generation—thirty years. Would you take it? No. Why? Because you are human and blunderers, of which I am one.

"Suppose, lastly, that the decision was made for you. Would you be happy? I doubt it. Stupidity, or human kindness, if you like, is the one thing that distinguishes us from that brainless reptile which Dr. Crane just showed us. For to be stupid is to be kind, and to be kind is to be stupid. Do not think I am bandying epigrams. I am not. Reflect. Is it not true—humanly true—that every time any one of you has given way to a decent 'human' impulse, he has kicked himself later for having been a fool? Think of it: kindness equals stupidity; stupidity equals kindness. If you doubt it, what keeps your hospitals for paupers full, your homes for the aged prosperous, and your institutes for incurables jammed to the doors? Reflect, I say; if you wish individuals to persist, when they should have perished; if you wish the race to perish, when it might persist, be kind, scorn intelligence, and choose an evolution which will send you back to the reptiles. I personally have no choice. Either alternative is a 'tale told by an idiot, full of sound and fury, signifying nothing.'

"But ladies and gentlemen, your decision has already been made. Four weeks from today you will know what has been decided for you. It is neither reptile nor superman, neither back to the brutes nor on to the gods. In the meantime—"

Certain light-witted members of the audience whose attention was already wandering, noticed Doctor Brown hurriedly follow a page off the platform and disappear behind the purple curtains. Joyously anticipating that the she-reptile had bitten and killed one of her keepers, they sat back, waiting for the chairman to reappear and announce the welcome tragedy. They were disappointed. Within a minute Brown was back, but on the floor of the auditorium. They saw him pluck the ranting speaker by the sleeve. But they did not hear what the Doctor whispered in De Soto's ear.

"Come with me at once. Alice—"

They were in a taxi before the audience realized that it was deserted. The second extras speculated on the significance of a scientific meeting—especially one of this importance—being abandoned without ceremony by speaker and chairman, but they drew no rational conclusions.

By noon the next day a verbatim account of that historic meeting, with the word for word reproduction of De Soto's speech, was printed in heavy type on the front page of every important newspaper or journal of the civilized world.

De Soto's son was born an hour after the meeting broke up. An hour later Alice was dead. Thirty minutes after Alice died, Brown reeled into Crane's apartment.

"She is dead, thank God! It was born alive."

"What is it?"

"I don't know. It is not a mammal. It is still alive. De Soto has it."

13

HIS LAST WILL AND TESTAMENT

POOR OLD WILSON HAD NOT PROSPERED SINCE his lodger deserted him. The all but deaf, half blind old man had puttered about for months in his inefficient endeavors to find a successor for the departed Bork, but without any luck. Finally he abandoned the effort to rent his shabby room, and resigned himself to fare a little more Spartan than what he had been accustomed to, in the days of his poor luxury. Bacon no longer was a possibility. Flapjacks, potatoes and the scanty greens from his own garden kept his sleeping old soul in his lethargic body. The neighbors became alarmed lest the old man die alone and thus smirch their shabby, decent street with the scandal of man's inhumanity to man.

One morning, long after his needy friends had given up all hope that old Wilson would ever again be on speaking terms with bacon and prosperity, he ambled proudly over to his nearest neighbor's with the glad tidings that the room was at last rented.

"Who to?" the gossip bawled in Wilson's better ear.

"Eh? I'm hard of hearing."

"Who did you rent the room to?"

"Durned if I know. Some fool. He paid in advance. Say,

is that a ten-dollar bill? My eyes ain't what they was."

Being assured that it was, old Wilson doddered off to the corner grocery and purchased a whole side of bacon. For four weeks the neighbors saw nothing of old Wilson or of his new lodger, although they kept a sharp watch for the latter. Finally the theory that the new lodger was a night worker, leaving the house after dark and returning before dawn, was generally accepted. Substantially it was true. The lodger left his room only between the hours of midnight and three in the morning, to purchase at cheap lunch counters and liquor joints the necessities of life. Not only was the shabby neighborhood totally unprepared for the tragedy which suddenly burst upon it, but also the whole world. At the end of the fourth week, when old Wilson began fretting lest his invisible lodger overlook the vital matter of the rent, the horror happened, without the slightest warning, at midnight of the last day of the fourth week. Fifteen minutes after it happened the tragedy was broadcast by telegraph, cable and radio to the farthest corners of the civilized world. And it was broadcast barely in time to save the human race from a similar fate.

The police were on the spot five minutes after the first inhuman scream shattered the dog-tired silence of the mean neighborhood's midnight. Even old Wilson heard that cry from hell. The siren of the police car shrieking through the night was not more shrill. The officers battered down the door of the lodger's room just as the last sounds of agony expired in a dying groan. Entering with drawn revolvers, they stumbled over a litter of empty bottles, dirty papers and fragments of half-eaten meals. The man had died defending himself against terrible odds. When they saw what had destroyed the victim, they froze where they stood. One officer recovered his senses and raised his arm to take aim. The captain knocked the automatic aside and the volley of shots went wild.

"Don't shoot! Get the envelope in his hand."

At the risk of his life the officer darted forward, snatched the envelope from the dead hand, and followed his shaking companion from the room.

"Get all the furniture in the house and block up the doorway!" the captain shouted. "Rip off the doors downstairs and bring them up. I'll stand guard."

While his men tumbled downstairs to fetch everything heavy that the shack contained, the captain glanced at the letter. It was stamped and addressed to Dr. Andrew Crane at the Erickson Foundation. Across the envelope "Private and Personal" was scrawled in red ink.

"Get this man on the telephone and tell him to come here at once," the captain ordered the first man who staggered up under a load of furniture.

Crane was located at his own address. The officer repeated the street and number. "Come at once."

"I'll be there in three minutes." Crane was as good as his promise. "Bork's old place," he muttered as he gave his machine the gas and shot into the street. "What now?"

He soon learned. The captain handed him the envelope and ordered him to read it. The meat of the letter was on the first page. Crane read it at a glance. Not bothering to look at the rest of the bulky manuscript or to inquire what had brought the police to Bork's old lodging, he stuffed the letter into his pocket and bolted for the stairs.

"Come back!" the captain shouted.

"Can't. This must be broadcast at once."

"Halt!"

The front door slammed after Crane just as a bullet flattened itself on the brass doorknob. He was roaring up the street before the second shot overtook him, missed him by an inch, and shattered the windshield of his low, open car. Shooting round a corner he put his pursuers hopelessly out of the running. Four minutes later he was seated before the broadcasting keyboard which the experts of the Erickson Foundation had especially designed at De Soto's suggestion for a purpose totally different from that for which Crane now used it.

Within half an hour his short, insistent message had girdled the globe. Newspaper broadcasting stations in those countries where it was still daylight or early evening took up the desperate message and drowned out all programs and other unnecessary interference. Where the radio failed, cables and wireless got drowsy engineers, sleeping editors and snoring politicians out of bed at unearthy hours, from Senegal to Capetown, from Shanghai to Valparaiso.

In the less sophisticated countries fire sirens shrieked through the streets in the dead of night; criers followed them up, yelling to the startled people to rise at once and destroy the plants, root and branch, of the newly constructed stations of the Universal Power Transmission Corporation. They needed no urging. De Soto's wild speech four weeks previously had been translated into every living language, popularized, and made accessible to the people of all countries. The telephotographs of Bertha's brood were in every newspaper office of the world two days after Wilkes had exhibited one of the seventeen living monsters that were now world-famous. Crane's selling campaign had been ably engineered for him before he ever sat down at the Erickson keyboard. The human race, or at least the civilized part of it, was already prepared for the hell about to burst upon it, which one man, who might easily have been shot by a stupid police officer, averted. Eight hours later the damage would have been beyond human repair.

Martial law in the more civilized countries made a feeble, ineffectual attempt to hold the raging mobs in control. Many were shot down in cold blood by the rifles and machine guns of the militias that vainly struggled to protect the property of an alien corporation. Property, to the custodians of human life, is more sacred than human life itself. Capital had been heavily invested in their bailiwicks. What did it matter what the scientists said? Money is money; business is business; and a fool is a fool the wide world over. Therefore the devoted members of the militia were butchered by sheer weight of numbers before humanity prevailed. For the argument of the people in this instance was beyond political or international palliatives. The deepest, fiercest instinct of the human race was about to be violated.

Instinct fought; civilization for the moment went under. From Senegal to Capetown, from London to Leningrad, from Shanghai to Valparaiso, from New York to San Francisco, and in every corrugated iron settlement of the earth's wilderness, a sombre torch of destruction flared up against the midnight skies or darkened the silver glare of the age-weary, tolerant sun. The vast plants of the Universal Power Transmission Company were destroyed the world over by flames and bombs four hours after Crane broadcast the first call to arms. Universal and all of its backers were ruined.

Thus far had the Erickson triumphed, but not in the way De Soto had predicted. What of the counter-attack De Soto prepared for the trustees? Crane destroyed that also. Before daybreak both the Erickson and the Universal were a total loss. One was destroyed outright with all of its equipment and all of its expensively captured markets; the other receded at one step to its comparatively harmless monopolies which it had held before De Soto tempted it with a vision of all the kingdoms of the earth. And what, through all this, of De Soto, the unprecedented world genius who had precipitated it all?"

To appreciate De Soto's motives, historians must take account of his own tragedy. Brown would have delivered him from the worst at the last moment—when it was too late. Alice should never have been permitted to bear a son to the husband she loved. This Brown admitted—when the son was born. There is a simple surgical trick, a quick snip of a pair of scissors, which is permitted in such circumstances to even the most conservative obstetrician. Brown would have used this, but the father forbade.

"Will she live?" he asked, referring to Alice.

"Only a few hours, at most."

"Then I refuse to have this thing put out of the world. It is mine. My first intention was right. So far as I am concerned, Alice is already dead. This episode in my life is ended."

De Soto was not with his wife when she died. He had already fled the house, taking with him his new-born son wrapped in a quilt.

For four weeks the world speculated on the fate of its greatest inventor, Miguel De Soto. Gradually the theory was accepted that he had destroyed himself in the sudden madness of grief when his beautiful young wife died. Brown did this much for the principals in the tragedy which he might have aborted: He signed a death certificate for the mother and stated that the father had the child. Alice's remains were cremated within thirty hours. Only Kent and the doctor witnessed the last. Brown of course told the heartbroken father that Alice had died naturally—as, indeed, she had. Nature, however, is hell.

The trustees of the Erickson Foundation mourned their brilliant director for two days. Then, convinced like the rest that De Soto had committed suicide, they reverently forgot him in a bronze tablet in the president's office, inscribed to "Miguel De Soto, Benefactor of Humanity and Founder of Our Fortunes." Finally, they decided not to canonize their Aladdin for twenty-six days, until the second phrase of their inscription would be an overwhelming fact.

As the days passed and no trace of De Soto was discovered by the police, Crane and Wilkes agreed with Brown that the unhappy father had indeed destroyed not only himself but also his offspring. Had he been alive, they argued, he must certainly have given some sign before his wife, whom he had loved, became an urnful of ashes. The three friends attended to the immediate present, and let the future go for the moment. De Soto's threat, that the world within four weeks would begin to solve its greatest problem, might be only the defiant gesture of a defeated maniac. They set about consolidating their definite scientific gains, writing up the voluminous report on the protozoa, giving the full history of Bertha and her reptilian brood, and finally putting forth the bold hypothesis that all of these apparent miracles were nothing more than the orderly progress of nature, hastened or retarded several billionfold by the control of radiation in relation to the germ cells of living animals.

Requests from every scientific center of the world for one of the artificially evolved—or rather, developed, reptiles poured in by the bushelful, and less presumptuous academies begged for at least one microphotograph of the perfected protozoa. The latter were easily satisfied. For months Wilkes had been preparing a new treatise, which was now published and sold as fast as the presses could print it. The more convincing proof, the seventeen living reptiles and their pickled baby brother, were started on a world tour two days after Alice died and De Soto disappeared. It is well that they did, for when it became necessary to destroy billions of dollars' worth of property, the public of at least one continent was thoroughly educated visually—and the world at large had seen hundreds of photographs of its grandparents.

The world was educated in one detail. When Crane began broadcasting the warning that unless the people of all countries at once destroyed the plants of the Universal Power Transmission Company, their own *children,* not possibly their great, great grandchildren, would be very similar to those reptilian grandparents now touring the civilized world, the warning struck home at once. Half an hour of wireless transmission by means of the new devices, Crane asserted, would suffice to change the germ cells of every living human being permanently. Thirty minutes, no more, he declared from the Erickson keyboard, would hurl every child born of parents then living back to the reptiles. Mothers would bear, not snakes, but things with legs and gigantic heads like those which the hen brought forth. These, however, unlike the hen's, would be born alive and not from the egg. At one stride the race would retrogress hundreds of millions of years to its premammalian ancestors. This, it was broadcast, would be the inevitable outcome of the first use of the new "Universal" system for the wireless transmission of electrical energy. The unborn would be born reptiles; the fruit of every union not yet consummated, for as long as the present generation lived, would be a race of carnivorous reptiles, possibly venomous.

The preservation of the species is a deeper instinct, even with the individual, than is the preservation of self. Bertha's fellow hens pecked her to death when their instincts taught them that she had betrayed the birds to the reptiles. Likwise when Crane, desperately transmitting De Soto's unintended warning from the Erickson broadcasting keyboards, spelled out the impending degeneration of the human race, instinct prevailed. Machine guns, gas and tear bombs, flame-projectors and human militia melted like smoke before a hurricane, when a race about to be outraged surged over the merely human defenses created by unlimited wealth. In four hours Universal was wiped out; the race was saved.

Money dies hard—if it ever dies. Crane, still sweating over the switchboards at nine in the morning, suddenly had a vision. Business, he saw in one transcendent flash of revelation, is immortal. Man may have a soul, the race a purposeful mind, but lucre has a belly and it has all-consuming powers of digestion and assimilation.

For seven hours the night and early morning had been turned into clanging day. Extra after extra headlined the progress of the world riots to excited householders as Universal's gigantic plants went their predestined way by bombs and flames. One shrewd go-getter after another hugged himself in the chilly dawn. These keen men of affairs had backed De Soto and the Erickson. Their money was safe. Had the Erickson been bombed? Not on your life! Were its trustees panicky? Again, not on your life. They knew what they were talking about when they cautiously released a "preliminary announcement" two days before the present fiasco of Universal. The canny trustees had presaged the collapse of the Universal. They themselves, they hinted, would broadcast the story of an invention which would scrap Universal in half an hour. Was this the prophesied revelation?

It was not. For two hours the perspiring president of the Erickson had been trying to distract Crane's attention. Crane stuck to his job, methodically transmitting the whole of De Soto's last will and testament. Universal was already destroyed; what Crane now did was a labor of hate. He broadcast the truth as De Soto's twisted, infinitely clear mind had conceived it, in the vain hope that common sense might at last prevail. He was deceived.

Unable to restrain himself any longer, the president roughly brushed Crane's hands off the keyboard.

"What do you think you are doing?" he demanded, red in the face, the veins on his neck and temples swollen to the bursting point.

"Putting a crimp in you, if you want to know," Crane grinned. "Get out of my way; I've got to finish this."

Appeals to gratitude for past benefits received, threats of arrest, promises of any reasonable sum up to fifty million, tears—almost all of these were offered and rejected in the brief space of ten minutes while Crane rested and the president wallowed.

"The man was crazy," the president all but sobbed as his final argument. "You know as well as I do that he had been out of his head for months. We've bluffed off the Universal. The world is ours. Fifty million if you stop broadcasting. You shan't—."

"Steady!" Crane ordered. "Hinder me now and I'll—. Sit down! Wait there till I'm through." He reached for a heavy steel spanner which some careless workman had left near the keyboard. "They can hang me if they like, but I'll smash your skull like an egg with this if you interfere."

As De Soto's last will and testament filtered into space, the whole purpose of his insane life became brutally evident and coldly clear. According to his own account he had fully intended helping the race a hundred million years in its struggle toward perfection, when he first realized his own incomparable powers.

Then, as the strange decay which was ultimately to undo him began to steal through his cells, he foresaw the futility of any help; for in the end the whole race must perish or be mutated into another, not human. Why strive for its perfection? What are a billion years in the life of the universe, where galaxies measure their moments by the pulse beats that are the birth and the lingering extinction of a noble race, like that of the kingly reptiles? On such a scale the chronology of the mammals, and their puny human offshoot, are less than the tenth of a second. The reptiles vanished, leaving only the comparatively indestructible accidents of their bony frames in the hardened sands; the mammals must follow their predecessors into oblivion; the very stars of heaven crumble to dust or dissipate in futile heat, and the records of all life's struggle must in the end be smoothed out in eternal cold. Why strive? To what end? Only an idiot would say "for the greater glory of the human mind." The reptile mind forgot its glory before the first mammal gave milk to its feeble young.

Pessimism, black, irrefutable, and absolute, seems to have been De Soto's creed at this transient stage of his own evolution. The next stage—induced, as he declared, by a blunder on his own part, which initiated the degeneration of his clear seeing mind—began when an accident in his laboratory started his descent. At first his purpose was clear and rational. A race that must perish, or at best lose its individuality beyond all hope of past memory, was not worth any rational being's efforts toward perfection. The longer it struggled to attain the unattainable, the longer would be its agony of frustration. Therefore, in mercy, it should be destroyed. This was De Soto's first purpose before he degenerated.

He was not brutal; destruction should come in thirty years, swiftly, painlessly, mercifully, like the dawn. How? By universal sterilization of the human race. The physical means were simple; he had grasped them in the first hour of his study after leaving the library. Not X-rays, but shorter radiations, capable of affecting the most intimately complex cells of the human body, could easily be broadcast over the entire earth in a short morning. He would save humanity from itself by wiping it out, painlessly, in an hour. Scientists would speculate for thirty years on the cause of the universal sterility. Their speculations would end in death, complete, quiet and peaceful for the whole human race; and no last handful of sages, hundreds of millions of years hence, would be condemned to see their dwindling star die and their leprous planet freeze. Sterilization, complete and universal for the race of men—that was the one sane answer to the riddle of the ages. When he joined the Erickson Foundation, this great dream was De Soto's purpose.

To accomplish it, he declared in his will, he needed technical assistance—broadcasting stations over the whole world. These he could not command without financial aid. Seeking that assistance, he met with his first doubt and his first check. Was he wrong after all? Is there something in mankind of a different order from anything that the splendid, perfect, all-conquering reptiles possessed? There was. Man, he learned to his astonishment, had a soul. Who knows, he sneered in his last will and testament, but that the carnivorous reptiles, who had two more or less centered nervous systems, had not a pair of souls? The human soul, De Soto declared in his will, shows itself in art.

To his perfectly adjusted nervous system, all human art appeared as a blundering attempt to harmonize what cannot be harmonized, and to seek proportion where none is possible. A certain restaurant inspired him to those reflections, and later, a cook and waiter induced him to apply similar principles to the human body and to its concomitant, the human mind. The soul of man, if it exists, is, he concluded, an abortion that should be chloroformed at birth. The reptiles, he asserted, had a better substitute.

De Soto was scientific. Although he scoffed at the existence of a human soul, he decided to experiment before declaring that it did not exist. Should such a thing be found, he would throw his unbounded talent aside and aid the race to develop this mysterious spark into a flame that would consume the universe. His will here becomes somewhat incoherent. In substance he seems to be saying that he offered mankind the stars and it asked for a better radio. Some who heard Crane's broadcasting of the original, interpret the obscure passage as meaning that De Soto offered all men everlasting oblivion, and they demanded eternal life. Whatever may be the correct interpretation, the objective facts are plain. De Soto was disgusted. In a last desperate battle he tried to educate a handful up to a taste for black pearls. His attempt failed, and he set out to make his friends kings and rulers of the world.

About this time, he asserts, he wavered. Might it not be possible, after all, to breed a race that would see nature eye to eye? How decide? Experiment answers all. He experimented, blundered, prove himself to be a human fool like all his kind, fell in love with his wife, tried to undo his blunder, failed, and, like the fool he admitted he was, doomed the whole race to follow in his own footsteps. He had hoped to show to all mankind, in his own son, an example of the transcendent genius that human nature, aided by human skill, may produce.

While experimenting he blundered—humanly, irrevocably. He grew to love his wife and longed for her death. Like the degenerated wretch he was, he could not kill her. He had failed. Was there still hope for the race? On a last appeal to the men whom he had made rich, he tried to make them see as he saw. The Universal, broadcasting its only half-understood wireless transmission of power, would avenge his own misfortune.

He knew that no living physicist or engineer could penetrate the subtle complexity of his mechanism. The best of them would see in it only a marvelously ingenious device for transmitting electrical energy without wires and without costly power stations. None would anlyze the inevitable consequences of the profitable transmission, for none had the inventor's all but superhuman genius. They could not calculate, as he did, from the subtle equations, the accompanying radiations that would spray the chromosomes of every human being with hard radiation. Before the keenest living physicist or biologist could suspect the danger, the damage would be done, and the whole race, profoundly changed in its most intimate germ cells, would be irrevocably reversed toward its reptilian ancestors. Like an explosion, the whole course of human, mammalian and later reptilian evolution would be undone in a single generation.

De Soto was not without mercy. Feeling that many might have at least the beginnings of a soul in their minds, he provided for them. An entire generation must bring forth only reptiles. This he had already ordained, in putting into Williams' hands the dangerous key to a financial fool's paradise. The wireless transmission of *electrical* energy, and with it the instantaneous pulse of *dysgenic* energy, degrading the unborn offspring of all then living to the outward shape and the inner bi-souled status of prehistoric reptiles, was a certainty. Williams and his crowd, human as fish, had swallowed hook, bait and sinker. They, not De Soto, should have the honor of hurling humanity backward hundreds of millions of years in one generation. So much for justice; mercy must be heard. De Soto's mercy was this, and it was adequate.

According to his last will, he baited Williams with the wireless transmission of *electrical* energy, and this is a fact. The counterblast with which he planned to destroy Williams was not electrical. At one stride De Soto put electrical energy forever on the shelf. It became as obsolete as the fly-coach—or it would have so become, had not Crane threatened the president with that hefty steel spanner. Atomic energy was the bait dangled by De Soto before the trustees' bulging eyes. At will he could pass up or down the atomic scale, transmitting any element into any other, as a skilled harmonist modulates his compositions, and in the passage from one element to its neighbors he released and controlled hells of energy that made the lightnings of heaven or the millions of volts dispensed by Universal as obsolete as the thin, steam whistle of a peanut stand. Many had released atomic energy; none had controlled it. De Soto did both, and he gave the great secret into the hands of the Erickson trustees as a free gift. They grasped it greedily. The moment Universal began marketing, the Erickson was to broadcast the full account of its own wireless "power"—controlled atomic energy—which would forever banish electricity and all its devices, as steam and gasoline had banished the plodding horse.

Included in that hard scale of cosmic rays, with which De Soto tempted and won the Erickson trustees, was another, a high harmonic of the first, tuned to disintegrate the procreative germ cells of all living things—plant, protozoon, animal, man. Thus would he show mercy. Universal's product would be on the air before the Erickson replied. For one generation the females of the human race would bring forth their reptilian young alive. Then, forever, the pulse of cosmic rays, generated from the disintegration of matter—universal matter, the stuff of which galaxies are made—disintegrated to swell the bellies of half a dozen human beings, would sterilize human and reptile alike. "Curtain," De Soto adds in his manuscript, "Humanity; Reptiles; Sterilization. The Great Comedy: Reptiles, Humanity, Extinction."

The last phrases of De Soto's great message to humanity flickered into space. The story of his own redemption by love, as his superb intellect rotted, is now a classic. Those who know it by heart may wonder why Alice was not redeemed by love, as De Soto was.

De Soto was not redeemed. He died as he lived. The letter which Crane broadcast was his last, futile gesture of triumph. The world was his, he said; the letter would lead the world through the hell it deserved for at least thirty years.

"You can broadcast this," the letter concluded, "as soon as you receive it, for then the Universal will have generated its reptiles in the bodies of your young women and in the cells of your young men. Nothing can ever again start evolution forward. We are reptiles, and as reptiles we shall live, propagate, and die, unless you accept my mercy. The Foundation which has taught me all that I know of humanity as it now is, may save at least the gray hairs of those now living from utter disgrace. Lest those whom I have served see their grandchildren—not merely their children—snapping at their prey, with reptilian bodies, I prescribe and offer you the solace of extinction. This, as I saw clearly at first, is the one hope of the human race. Use my device for the generation and transmission of atomic energy, and within a century the human race will have perished. For your offspring now will be reptiles; thereafter you and they will be sterilized—if you use my device. It will give you world supremacy in finance for as long as the human race endures.

"I mail you this when it is too late to avert the disaster which Universal will precipitate tomorrow. May the whole race taste the bitterness which I have drained. Once I had a vision, I blundered, I loved. I blundered again and again. I had a vision only to blunder irrevocably. The one that I loved is dead

—dead as the whole futile human race will some day be, and she has left me a son, the sum and substance of all my blunders. Like you that I despised and would once have helped, I am a failure, undone by my own humanity. I cannot hate you, for you are reptiles, even as I am. Your intellect, like mine, at its best, is no better than the blundering instincts of a thing that perished before the first of our kind was conceived. Why prolong the farce? For thirty years you will see yourselves as you were, are, and shall be. Then the curtain will drop forever on this silly interlude of eternity. You will find my body with my son's. His mother would love neither of us, could she see us now. 'As I am, you shall be.' "

"I guess you're wrecked too," Crane remarked to the president as he finished broadcasting De Soto's testament. "We all are. Shut up! Get out."

The president left hurriedly, and Crane called up the doctor.

"Come round to Bork's place with me, will you? I haven't the nerve to go alone."

Fifteen minutes later they were cautiously admitted by the police. Crane was now respectable—the extras had restored his good name. The captain volunteered to lead an expedition into the barricaded den.

"Let me go first," he advised. "It has been moving about and whistling birdlike for the last two hours."

The barricade was cautiously removed. Not a sound issued from the room. The living thing within had taken its brainless revenge on the author of its unnatural life. With steady hand the captain aimed for the reptile's rudimentary upper brain and pulled the trigger. Three short convulsive jerks, and the monstrous son expired in shambles which had been his father.

—THE END—

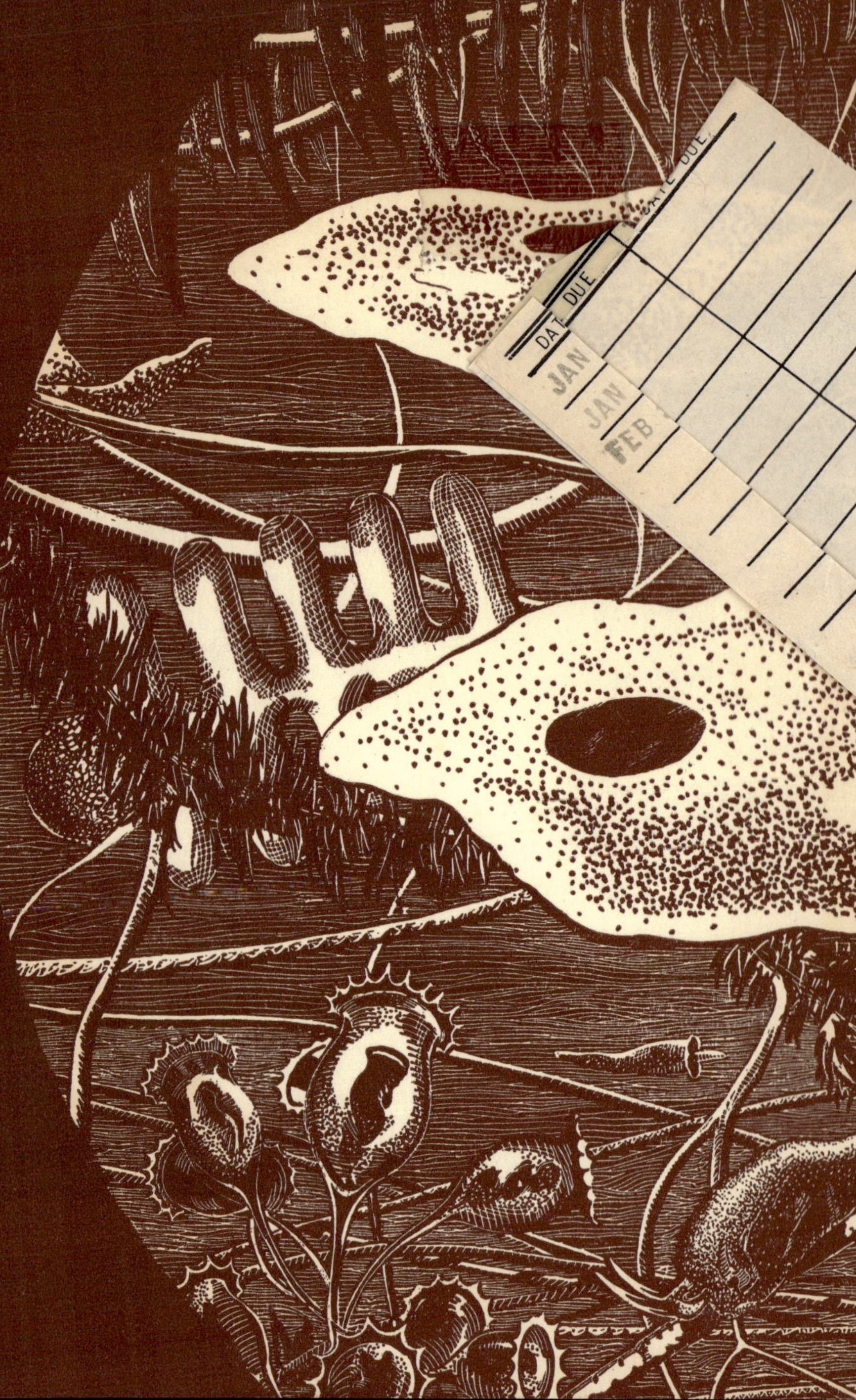